W0042168

Introduction to nuclear radiation detectors

INTRODUCTION TO

Nuclear Radiation Detectors

P.N.COOPER

Department of Electrical and Electronic Engineering and Applied Physics, Aston University, Birmingham

The right of the
University of Cambridge
to print and sell
all manner of books
was granted by
Henry VIII in 1534.
The University has printed
and published continuously
since 1584.

CAMBRIDGE UNIVERSITY PRESS

Cambridge

London   New York   New Rochelle

Melbourne   Sydney

CAMBRIDGE UNIVERSITY PRESS
Cambridge, New York, Melbourne, Madrid, Cape Town,
Singapore, São Paulo, Delhi, Tokyo, Mexico City

Cambridge University Press
The Edinburgh Building, Cambridge CB2 8RU, UK

Published in the United States of America by Cambridge University Press, New York

www.cambridge.org
Information on this title: www.cambridge.org/9780521281324

© Cambridge University Press 1986

This publication is in copyright. Subject to statutory exception
and to the provisions of relevant collective licensing agreements,
no reproduction of any part may take place without the written
permission of Cambridge University Press.

First published 1986
First paperback edition 2011

A catalogue record for this publication is available from the British Library

Library of Congress Cataloguing in Publication data
Cooper, P. N.
Introduction to nuclear radiation detectors.

Includes index.
1. Ionizing radiation—Instruments.   2. Nuclear counters.
1. Title.
QC795.5.C66   1986   539.7′7   86–2228

ISBN 978-0-521-26605-5 Hardback
ISBN 978-0-521-28132-4 Paperback

Cambridge University Press has no responsibility for the persistence or
accuracy of URLs for external or third-party internet websites referred to in
this publication, and does not guarantee that any content on such websites is,
or will remain, accurate or appropriate.

Contents

··· ···

Preface

………………………………………………………………………………

This book is intended primarily as a textbook for use in first degree courses. It is designed to be a stand alone book and no extensive pre-knowledge of nuclear physics is assumed. To this end an introduction to the main types of ionising radiations that may be encountered is included, as well as a summary of counting statistics. The main detector types that will be found in laboratory and project work are covered and the principles of the most common electronic processing modules are explained. Detailed circuits are not given since there is a wide selection of commercial high performance electronic instruments available. A chapter on the measurement of doses and dose rates of ionising radiations is also included which will make the book of interest to health physicists and safety officers.

P. N. Cooper

1

Ionising radiations

1.1 Radioactive decay

Ionising radiations most commonly arise from the decay of radioactive nuclei. An atom consists of a nucleus containing protons and neutrons surrounded by a cloud of electrons. The nucleus is small, of the order of 10^{-4} of the overall diameter of the atom, which is a few times 10^{-10} m in diameter. The nucleus, however, contains nearly all of the mass of the atom. Protons and neutrons each have a mass of about 1.67×10^{-27} kg (the neutron is marginally heavier than the proton) and the electrons have a mass of 9.11×10^{-31} kg. Neutrons are uncharged and protons and electrons carry charges of $+1.6 \times 10^{-19}$ C and -1.6×10^{-19} C respectively. In a neutral atom the number of electrons will therefore equal the number of protons in the nucleus. This number is conventionally called the *atomic number* and is denoted by the symbol Z. The total number of neutrons and protons in the nucleus is represented by the *mass number*, A, which is very close to the atomic mass measured on the *mass unit* scale. One mass unit (1 u) is 1.66×10^{-27} kg and is defined as being exactly $\frac{1}{12}$ of the mass of the $^{12}_{6}C$ atom. The superscript to the chemical symbol denotes the mass number and the subscript denotes the atomic number. Two types of stable carbon atom exist naturally, $^{12}_{6}C$ and $^{13}_{6}C$ and radioactive $^{14}_{6}C$ is also found. In addition it is also possible to create another form of radioactive carbon $^{11}_{6}C$. All these types of carbon atom, with different mass numbers and hence different numbers of neutrons in the nucleus are called *isotopes* of carbon. These isotopes have identical chemical properties since these are controlled by the number of electrons and cannot be separated from each other by chemical means. Most elements have more than one stable isotope and the distribution of these isotopes, together with some of the longer lived naturally occurring isotopes is shown in Fig. 1.1. Also shown are some of the man-made elements with atomic numbers greater than that of uranium (92) that are consequently termed the *Transuranic elements*. This diagram, which shows the atomic number and neutron number of the isotopes is known as the Segre chart and all isotopes of the same element occur on the same horizontal row.

Comparison of the value of the mass unit with the masses of protons and neutrons shows that since the isotopic masses are all within 0.1 u of the mass number then all individual atoms have a slightly lower mass than the sum of the masses of their constituent protons, neutrons and electrons. When the atoms were formed, a small amount of mass, known as the mass defect, was converted into energy. The mass defect is just the difference between the mass of the component parts and the mass of the finished atom.

Fig. 1.1. Segre Chart. The atomic number (protons) versus neutron number for stable isotopes and longer lived natural radioactive and transuranic elements.

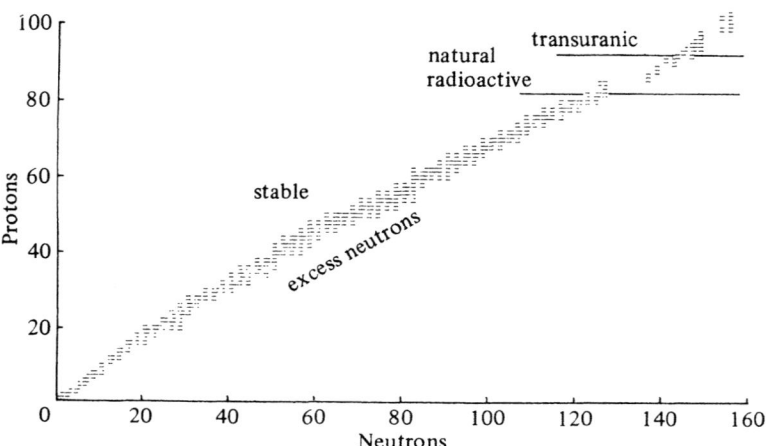

The relation between the change in mass Δm and energy released is given by the Einstein relation $E = \Delta mc^2$, where c is the velocity of light in free space (3×10^8 m s^{-1}). The energy released on formation of an atom from its component parts is also equal to the energy required to completely break up the atom and is known as the binding energy. This binding energy is not linearly proportional to the mass number and the variation is more clearly shown by dividing the binding energy by the mass number to obtain the average binding energy per nucleon. Nucleon is a name that covers both protons and neutrons within the nucleus. Atoms that are most stable have the highest value of average binding energy per nucleon. In Fig. 1.2. the negative of the average binding energy per nucleon, here termed the potential energy, is plotted against mass number and the most stable atoms have the lowest potential energy in this diagram.

The unit of energy most commonly used in connection with the nucleus is the mega electron volt, or MeV. An energy of 1 MeV is given to a particle of unit charge (1.6×10^{-19} C) when it is accelerated from rest across a potential difference of one million

Fig. 1.2. Potential energy per nucleon versus mass number.

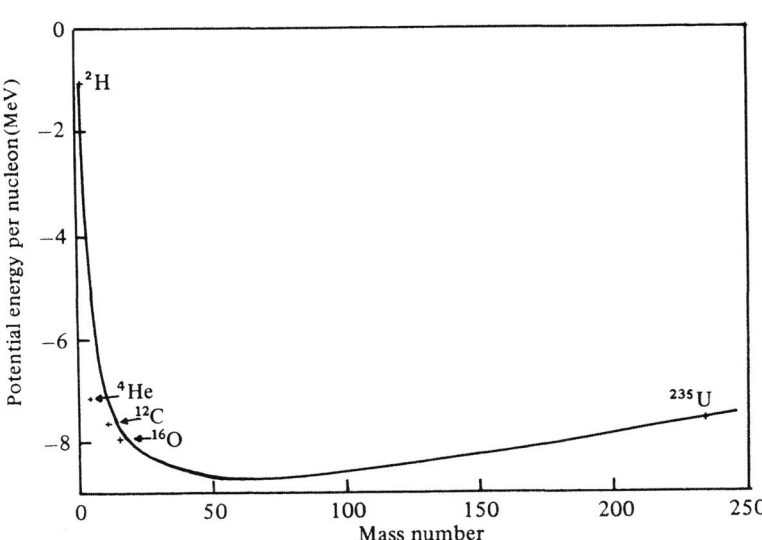

volts, and is equal to 1.6×10^{-13} J. Energies of X-rays are generally expressed in keV where 1 keV is equal to 1.6×10^{-16} J and the energies of electrons taking part in chemical bonding (valence electrons) are generally expressed in electron volts (eV) where 1 eV is equal to 1.6×10^{-19} J.

Isotopes that do not lie within the narrow band of stability shown in Fig. 1.1 have a smaller binding energy than stable isotopes of the same mass number and will try to reach a condition of greater stability by radioactive decay. For isotopes of $Z \geqslant 82$ which have excess neutrons the most common mode of decay is by emission of an ordinary negative electron, termed in radioactive decay a beta-particle. Electrons do not exist inside the nucleus but it is possible for the reaction

$$n \rightarrow p + e^-$$

to occur within the nucleus. The proton remains in the nucleus but the electron or beta-particle is promptly ejected, taking some energy with it. In beta decay the mass number A is unchanged but the atomic number Z is increased by one, thus producing an isotope of the element one position higher in the periodic table. One example of such a decay is

$$^{14}_{6}C \rightarrow {}^{14}_{7}N + \beta^- + \bar{\nu} + 0.156 \text{ MeV}$$

The time taken for half of the initial number of radioactive atoms to decay, which is known as the half-life, is 5730 years. The extra particle emitted in this reaction is the anti-neutrino, denoted by the symbol $\bar{\nu}$. The anti-neutrino has no charge, a mass very small compared with that of the electron and its purpose is to conserve the intrinsic spin of the particles involved in the decay.

If, however, isotopes having an excess of protons compared with stable isotopes of the same element are considered it is found that a different mode of decay exists, the most common mode of decay being the emission of a positively charged electron, commonly known as the positron, which is identical to the normal negative electron except for the sign of its charge. One example of such a decay is

$$^{11}_{6}C \rightarrow {}^{11}_{5}B + \beta^+ + \nu + 0.96 \text{ MeV}$$

and the half-life for this decay is 20.4 minutes. The extra particle emitted in this decay is the neutrino, denoted by the symbol ν, which is virtually identical to the anti-neutrino but is regarded as its anti-particle, an unimportant distinction for most decays.

A very important consequence of the emission of neutrinos in positron decay and anti-neutrinos in beta decay is in the energy of the emitted positive or negative electron (β-particle) since the energy released is shared by three particles. As a result the β-particles can have any energy in the range zero to the energy released in the reaction (less a very small correction for the energy carried away by the recoiling nucleus). The general shape of the energy spectrum of β-particles emitted from a radioactive source is shown in Fig. 1.3. Note that in both β and β^+ decays the mass number is unchanged but the atomic number changes by $+1$ and -1 respectively.

For certain isotopes with mass number greater than or equal to 82 (lead), decay occurs by emission of a ^4_2He nucleus, which is known as the alpha-particle. The alpha-particle has a mass of

Fig. 1.3. Typical energy spectrum of beta-particles emitted from a radioactive source.

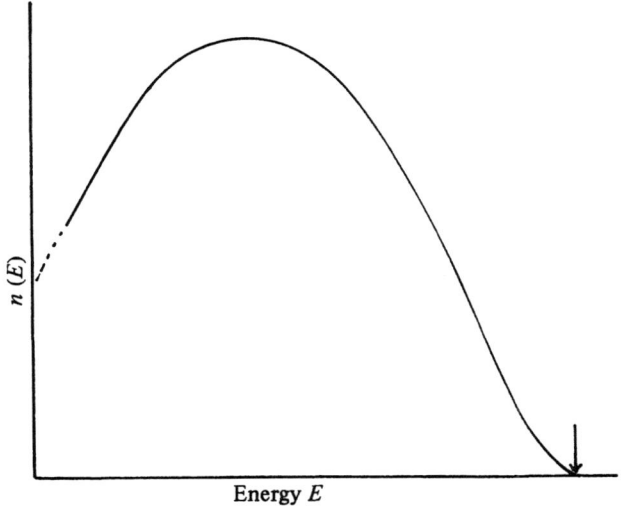

Energy E

6.64×10^{-27} kg and is doubly charged since it is a strongly bound combination of two protons and two neutrons. An example of one such decay is

$$^{210}_{84}\text{Po} \rightarrow {}^{206}_{82}\text{Pb} + \alpha + 5.4 \text{ MeV}$$

This decay has a half-life of 138 days and all alpha-particle decays, unlike beta decays, result in only two particles, the alpha-particle and the recoil nucleus. The energy is therefore shared uniquely between these two particles in the inverse ratio of their masses, a result easily proved from conservation of energy and momentum. In the above example the alpha-particle will take $\frac{206}{210}$ of the total energy (5.297 MeV) and the recoiling ^{206}Pb nucleus will take $\frac{4}{210}$ of the total energy (0.103 MeV). It will be noticed that in alpha-particle decays the mass number decreases by 4 and the atomic number by 2.

The naturally occurring radioactive isotopes at the top end of the periodic table can be grouped into three decay chains that are a mixed sequence of alpha-particle and beta-particle decays with widely differing half-lives. These chains are:

  (a) Thorium series ($A = 4n$) starting with ^{232}Th which is an alpha-particle emitter of half-life 1.39×10^{10} years and ending with stable ^{208}Pb.
  (b) Uranium series ($A = 4n + 2$) starting with ^{238}U which is an alpha-particle emitter of half-life 4.5×10^9 years and ending with stable ^{206}Pb.
  (c) Actinium series ($A = 4n + 3$) starting with ^{235}U which is an alpha-particle emitter of half-life 7.07×10^8 years and ending with stable ^{207}Pb.

It will be noticed that one series ($A = 4n + 1$) is missing, that is, it does not occur naturally. The missing series consists of radioactive isotopes which all have half-lives much shorter than the age of the earth (4.5×10^9 yr) and so, although the chain would have been present after the formation of the earth it has now decayed to unmeasurably small proportions. The members of the chain have, however, been synthesized by either neutron or charged particle bombardment of heavy isotopes. It is known as the Neptunium series and starts with the longest lived member of the chain, ^{237}Np, which is an alpha-particle emitter of half-life 2.2×10^6 years, and ends with stable ^{209}Bi.

In most radioactive decay reactions the final, or daughter, nucleus is commonly left in an excited state from which it decays to the most stable state, or ground state, by emission of one or more quanta of electromagnetic radiation, known as gamma-rays. There is no real distinction between gamma-rays and X-rays except in their modes of production, and the energy ranges of gamma-rays and X-rays have a wide overlap. The energy of a single quantum of gamma-radiation may range in energy from about 10 keV to several MeV and the wavelength is given by

$$\lambda = hc/E$$

where h is Planck's constant (6.62×10^{-34} Js) and c is the velocity of light in free space ($3 \times 10^8 \, \mathrm{m\,s^{-1}}$). The wavelength λ is expressed in metres when the energy E is in joules. Since, as already given, 1 MeV $= 1.6 \times 10^{-13}$ J, then the wavelength can be expressed as

$$\lambda = (1.24 \times 10^{-12})/E \text{ m}$$

when the energy E is given in MeV. The examples of decays already given are rather rare in that they do not emit any gamma-radiation. Some examples of particulate decay accompanied by gamma-radiation are:

(a) ^{60}Co $\rightarrow$ ^{60}Ni $+ \beta^- + \bar{\nu} + 2\gamma + 3.99$ MeV (5.3 yr half-life) where the beta decay takes 1.49 MeV and the gamma-rays have energies of 1.33 MeV and 1.17 MeV.

(b) ^{22}Na $\rightarrow$ ^{22}Ne $+ \beta^+ + \nu + \gamma + 1.82$ MeV (2.6 yr half-life) where the positron decay takes 0.55 MeV and the single gamma-ray has an energy of 1.27 MeV.

(c) ^{226}Ra $\rightarrow$ ^{222}Rn $+ \alpha + \gamma + 4.867$ MeV (1600 yr half-life). This decay occurs in 5.5% of the disintegrations, giving a gamma-ray of 0.186 MeV and an alpha-particle of 4.598 MeV. In the remaining 94.5% of the decays no gamma-radiation is emitted and the alpha-particle has an energy of 4.781 MeV with the recoiling nucleus of ^{222}Rn taking the remaining 0.086 MeV. The gamma-ray spectrum from an actual source of ^{226}Ra is in reality much more complex than indicated above since the shorter lived decay products build up and these have their own gamma-ray emissions that add to that of the parent nucleus.

1.2   Ionising effects of nuclear radiations

Any charged particle is capable of causing ionisation when it passes through gaseous, liquid or solid matter. The amount of energy lost by the charged particle in any individual ionising event shows very large fluctuations, but it must exceed the energy needed to remove an electron from an atom of the stopping material (the ionisation potential) or the energy needed to excite an electron into an empty energy level in a solid such as a semiconductor. An approximate value for the average energy needed to create an ionisation in a gas is 35 eV and is a figure that can be applied to most gases despite their having different ionisation potentials. In semiconductors an ionising event does not produce a positive ion and an electron as in a gas, but produces an electron and a positive hole. A positive hole is in reality the absence of an electron from a normally electrically neutral crystalline lattice composed of fixed positive ions and moving electrons. Positive holes move by an electron filling the vacancy and so moving the hole to the previous position of the electron. It is, however, convenient to regard positive holes as if they were real particles, but they must never be confused with positrons. In silicon an average expenditure of 3.6 eV is required to create a hole–electron pair (energy gap is 1.1 eV) and in germanium only 3.0 eV on average is required (energy gap is 0.7 eV).

1.3   Interaction of charged particles with matter

The linear stopping power or specific energy loss in a given absorber is defined as the rate of loss of energy per unit distance:

$$s_x = -(\mathrm{d}E/\mathrm{d}x)$$

and varies with energy and hence the distance x. If the particle has a fixed charge ze where e is 1.6×10^{-19}C and it is travelling through an absorber of atomic number Z which has N atoms per unit volume then the linear stopping power is given by the Bethe formula:

$$\frac{-\mathrm{d}E}{\mathrm{d}x} = \left(\frac{1}{4\pi\varepsilon_0}\right)^2 \frac{4\pi e^4 z^2}{m_0 v^2} NB \; \mathrm{J\,m^{-1}} \tag{1.1}$$

where

$$B = Z\left[\ln\frac{2m_0 v^2}{I} - \ln\left(1 - \frac{v^2}{c^2}\right) - \frac{v^2}{c^2}\right] \tag{1.1a}$$

In the above equations (1.1) m_0 is the mass of the electron and I is the average excitation and ionisation potential of the absorber. This expression fails at low particle energies or velocities since the charge on the ionisation particle ze decreases. For example, an alpha-particle may become singly charged for a few collisions and then may regain double charge at a further collision, the average charge decreasing as the velocity decreases. In a similar manner a singly charged particle may temporarily become neutral and also have an average charge that decreases with decreasing velocity. The overall effect is of a decreasing and effectively non-integral value of z near the end of the particle's path.

Since ionisation is a statistical effect the distance travelled by similar particles having the same initial energy before they finally become neutral shows fluctuations, or straggle, about the mean value, known as the mean range of the particle. A single particle will have a specific energy loss in a uniform homogeneous absorber, as shown in Fig. 1.4 for a typical alpha-particle in air. The effect of decreasing effective charge as the particle nears the end of its range can clearly be seen. In Fig. 1.5 the straggle in range observed for observations on a large number of particles can be seen. The foregoing discussion only applies to heavy particles such as alpha-particles and protons that will undergo negligible change in direction at each ionising interaction and will therefore travel in a straight line from the source. A useful figure for detector design is the mean range of such heavy charged particles as a function of their initial energy. Fig. 1.6 shows the range energy relationship for

Fig. 1.4. Specific energy loss for a typical alpha-particle in air.

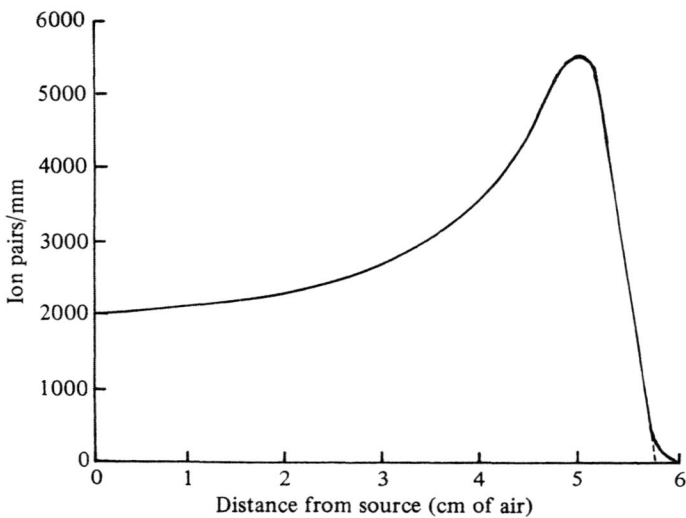

Fig. 1.5. Typical straggle in range for alpha-particles in air.

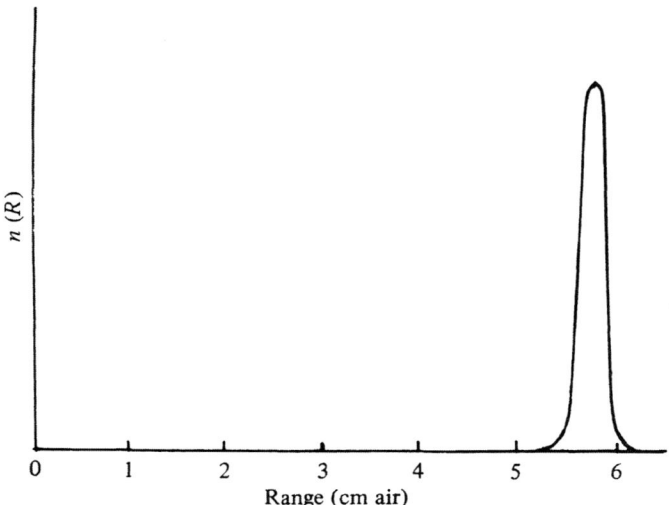

protons, deuterons and alpha-particles in silicon, and in Fig. 1.7 similar information is given for a widely used plastic scintillator, NE102A (see Chapter 5).

Fast electrons, being very light (only $\frac{1}{1832}$ of the proton mass) are easily scattered and follow a meandering path during slowing down. The specific energy loss due to ionisation and excitation is given by an expression somewhat similar to equation (1.1), but the electrons can also lose a significant amount of energy by radiative processes. The emitted radiation is known as *bremsstrahlung*

Fig. 1.6. Range–energy relationship for protons, deuterons and alpha-particles in silicon. (Data from C. F. Williamson, J. B. Boujot and J. Picard, CEA-R-3042 (July 1966).)

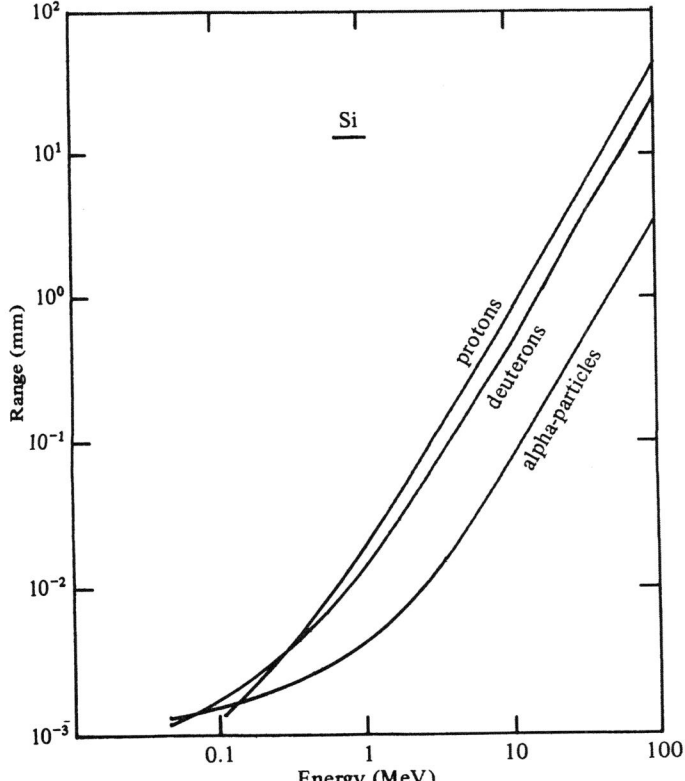

(braking radiation) and is the source of the continuous X-ray distribution from high voltage X-ray tubes. Production of bremsstrahlung gives an additional specific energy loss roughly proportional to the energy of the electron and the z^2 of the stopping material. Even for high Z absorbers the most significant energy loss is still by ionisation and excitation.

Because of the meandering path of electrons, even mono-energetic electrons in a collimated beam do not have an exact straight line range from the source, although an average range can be defined that is less than the total path length travelled by the

Fig. 1.7. Range–energy relationship for electrons, protons, deuterons and alpha-particles in plastic scintillator NE102A. (Reproduced by permission of Nuclear Enterprises Ltd, Edinburgh.)

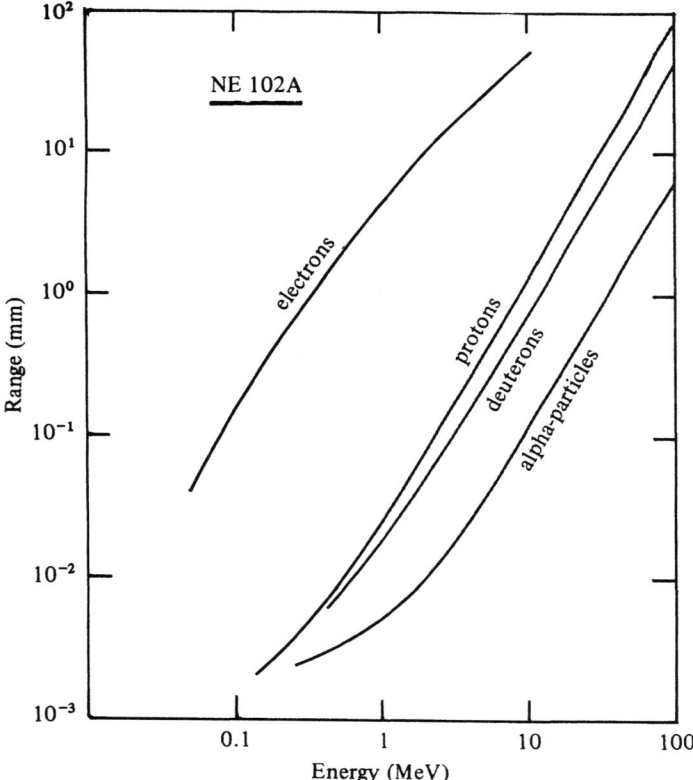

electron. An effective range energy curve for electrons in aluminium is shown in Fig. 1.8 where the range is expressed in terms of mass per unit area of absorber, which is linear range times density.

For electrons from a radioactive source (β-particles) the range energy relationship is complicated by the continuous energy distribution from zero up to the end point energy. A plot of the number of electrons reaching a given distance from the source that is less than the maximum range of the highest energy electron has a roughly exponential decrease with increasing distance, but this obviously only holds up to the maximum electron range present. Some residual ionisation can be detected beyond this point but this is due to production of bremsstrahlung by the slowing down electrons which will penetrate beyond the maximum range of the charged particles. The effect is more obvious for high Z absorbers.

Fig. 1.8. Range–energy relationship for electrons in aluminium. (Reproduced by permission of Canberra Instruments Ltd, Swindon.)

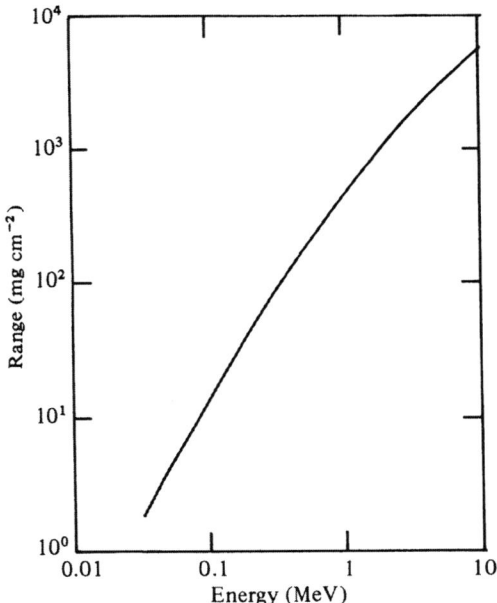

Since electrons are easily scattered in ionising collisions, then some will be backscattered towards the source from an absorber. For absorber layers that are thin compared with the maximum electron range the backscatter rate from β-particle emitting sources increases approximately linearly with increasing thickness of the scatterer and reaches a constant value for thick scatterers. Also the backscatter rate increases with increasing Z of the scatterer but, since it also depends to some extent on the number density of atoms in the scatterer, it is not a precise method for identification of elements. Entrance windows on detectors can therefore cause a reduction in the number of electrons detected compared with the number incident on the detector. The foregoing applies to both negative electrons and positrons from β-particle emitting sources, but a further effect is found with positrons which are anti-particles of electrons. When a positron reaches the end of its path it will annihilate with any convenient electron and so the mass of two electrons is turned into energy. This energy appears as two gamma ray quanta of 0.51 MeV energy each. Consequently shielding against positrons is more difficult than against negative electrons since the annihilation radiation requires additional shielding.

1.4   Gamma ray interactions with matter

Gamma-rays are uncharged electromagnetic quanta and consequently cannot cause ionisation directly. They can, however, interact with an absorber in a variety of ways, depending on their energy. The interactions of interest for radiation measurements are those that will produce fast electrons that in turn will cause ionisation in a similar manner to β-particles. The three interaction mechanisms are photoelectric absorption, Compton scattering and pair production and these will be described in turn.

1.4.1   Photoelectric absorption

In this interaction the gamma-ray quantum (or X-ray) interacts with a strongly bound electron in an absorber atom and completely disappears, producing a fast electron. Generally the gamma-ray will interact with the innermost, or K-shell of the atom provided that it is energetic enough to ionise the K-shell. The photoelectron is emitted with an energy less than the gamma-ray energy by the binding energy of the electron in the shell:

$$E_e = E_\gamma - E_K \qquad (1.2)$$

Since this process leaves a vacancy for an electron in the K-shell it may be filled by an electron from an outer shell, resulting in one or more quanta of X-radiation.

An alternative mode of de-excitation that is more probable in low Z absorbers is the emission of an Auger electron. In this process the excitation energy is transferred directly to one of the outer electrons, which is then emitted with an energy equal to the excitation energy less its own binding energy.

The photoelectric effect is stronger for lower energy gamma-rays and for absorber materials of high atomic number and the probability of the photoelectric interaction occurring is roughly proportional to

$$Z^n/E_\gamma^3$$

where n varies between 4 and 5. For this reason high atomic number materials are better for shielding purposes and for detection of gamma-rays, especially where energy measurements are to be made.

1.4.2 Compton scattering

Unlike the photoelectric effect, which can only occur with strongly bound electrons, Compton scattering is an interaction between gamma-rays or X-rays and free or only weakly bound electrons. The gamma-ray quantum transfers only part of its energy to the electron and is scattered at a lower energy in the process. In Fig. 1.9 the Compton scatter is depicted and the angles of scatter of the electron and gamma-ray are defined for use in the following equations. The energy of the scattered gamma-ray as a function of the angle of scatter θ is

$$E_\gamma' = \frac{E_\gamma}{1 + \dfrac{E_\gamma}{m_0 c^2}(1 - \cos\theta)} \qquad (1.3)$$

and, from conservation of energy,

$$E_\gamma = E_\gamma' + E_e \qquad (1.4)$$

where m_0c^2 is the rest mass energy equivalent of the electron, equal to 0.511 MeV. Except for very low incident energies the angular distribution of the scattered gamma-rays is strongly forward peaked. Two extreme values are important in Compton scattering. The first is the minimum energy of the scattered gamma-ray that occurs at a scattering angle $\theta = 180°$. This minimum energy gamma-ray is often called the backscattered gamma-ray and its energy, obtained by putting $\cos \theta = -1$ in equation (1.3), is given by

$$E_{\gamma BS}' = \frac{E_\gamma}{1 + \dfrac{2e_\gamma}{m_0c^2}} = \frac{m_0c^2/2}{1 + \dfrac{m_0c^2/2}{E_\gamma}} \qquad (1.5)$$

from which it can be seen that the energy of the backscattered gamma-ray cannot exceed 0.2555 MeV even for very high energy incident gamma-rays.

The second extreme value of interest is the energy of the electron recoiling at $\phi = 0°$, often termed the Compton edge energy. This forward scattered electron accompanies the 180° backscattered gamma-ray since there can be no transverse component of momen-

Fig. 1.9. Compton scattering of a gamma-ray or X-ray quantum.

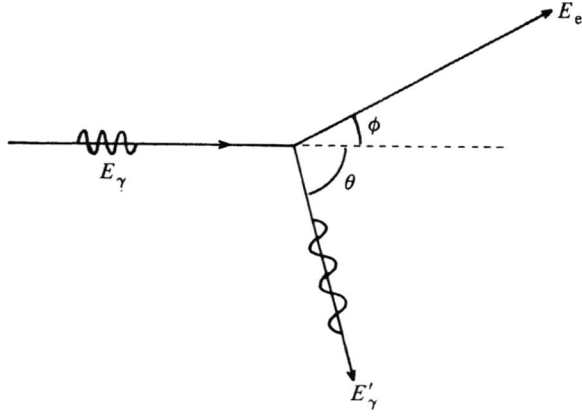

tum in this case and its energy is easily found from conservation of energy to be

$$E_{eCE} = \frac{E_\gamma}{1 + \dfrac{m_0c^2/2}{E_\gamma}} \tag{1.6}$$

From equation (1.6) it can be seen that the Compton edge energy is always less than the incident gamma-ray energy and for gamma-ray energies in the region of 1 MeV the difference is about 0.2 MeV.

For detection of gamma-radiation another important quantity is the probability of the scattered electron having a particular energy in the range zero to the Compton edge energy. This is expressed by a quantity termed the differential cross-section for electron scatter as a function of the energy of the scattered electron. If the incident gamma-ray energy and the scattered electron energy are expressed in the reduced units

$$\gamma = E_\gamma/m_0c^2 \quad \text{and} \quad t = E_e/m_0c^2$$

then the differential cross-section per electron is given by

$$\frac{d\sigma}{dE_e} = \frac{0.4892}{\gamma^2} \left[2 + \frac{t^2 - 2t}{\gamma^2 - \gamma t} + \frac{t^2}{(\gamma^2 - \gamma t)^2} \right] \text{ barns MeV}^{-1} \tag{1.7}$$

where the unit of cross-section, the *barn*, is equal to 10^{-24} cm^2. Equation (1.7) only applies over the energy range $0 \leq E_e \leq E_{eCE}$. By integration of equation (1.7) it is possible to obtain the total Compton cross-section per electron, the result being

$$\sigma = 0.500 \left[\frac{(1 + \gamma)}{\gamma^2} \left(\frac{2(1 + \gamma)}{(1 + 2\gamma)} - \frac{\ln(1 + 2\gamma)}{\gamma} \right) \right.$$
$$\left. + \frac{\ln(1 + 2\gamma)}{2\gamma} - \frac{(1 + 3\gamma)}{(1 + 2\gamma)^2} \right] \text{ barns} \tag{1.8}$$

The probability of Compton scattering depends upon the number of available (i.e. weakly bound) electrons as well as on the cross-section and so it increases with both the atomic number of the scatterer and the incident gamma-ray energy. Whereas the probability of Compton scattering for a carbon atom $(Z = 6)$ is almost exactly six times that for a hydrogen atom over a wide energy range, the probability for lead ($Z = 82$) is 26.3 times that of hydrogen at 0.01 MeV, 69.7 times at 0.1 MeV and 81.5 times at 1 MeV.

1.4.3   Pair production

This reaction, which results in the production of a positron and a negative electron is only energetically possible provided that the gamma-ray energy exceeds twice the rest mass energy equivalent of the electron or 1.022 MeV. It only has a low probability until the gamma-ray energy has about twice this value and so is mainly observed for higher energy gamma-radiation. In this interaction, which has to take place in the field of a nucleus, the gamma-ray completely disappears and an electron–positron pair is created, the two particles sharing the available energy of $E_\gamma - 1.022$ MeV. Just as for positrons released in radioactive decay the positron will, after being slowed down to rest, annihilate with the nearest available electron and so release two gamma-ray quanta of 0.511 MeV each.

Figs 5.2 and 5.3 in Chapter 5 show the relative contributions of the three gamma-ray absorption processes for two widely used scintillators, NaI and a plastic scintillator NE102A. The photoelectric effect is quite marked in NaI and is mainly due to the iodine which has a Z of 53. In the plastic scintillator, which is mainly composed of carbon and hydrogen, the photoelectric effect is very small and negligible for gamma-rays in the range from common radioactive emitters.

1.5   Neutrons

Neutrons, which have a mass similar to that of the proton but have zero charge, may be encountered with energies from thermal (of the order of kT) upwards depending on the method of production. A widely available neutron source for laboratory use contains a mixture of finely divided beryllium powder and an alpha-particle emitter. This produces neutrons by the following reaction:

$$^9\text{Be} + {}^4\text{He} \longrightarrow \begin{array}{l} {}^{12}\text{C} + \text{n} \\ {}^{12}\text{C*} + \text{n} \\ 3{}^4\text{He} + \text{n} \end{array}$$

Fast neutrons with energies up to 10 MeV can be obtained with an average energy of around 5 MeV but the spectrum of neutrons obtained depends on the alpha-particle emitter used and on the detailed physical construction of the source, particularly the fineness of the beryllium powder and the emitter and their degree of mixing. Fig. 1.10 shows a typical spectrum for an ^{241}Am–Be source. Since ^{241}Am has a half-life of 458 years it maintains a fairly constant output with time but suffers the disadvantage that the ^{241}Am does produce 60 keV gamma-rays as well as alpha-particles. All such sources, whatever the alpha-particle emitter, also produce high energy gamma-rays of 4.4 MeV from the decay of the excited state of ^{12}C (^{12}C*) to its ground state and the yield is roughly one gamma-ray for every three neutrons emitted.

Fig. 1.10. Typical neutron energy spectrum from an ^{241}Am–Be source measured with an NE213 scintillator.

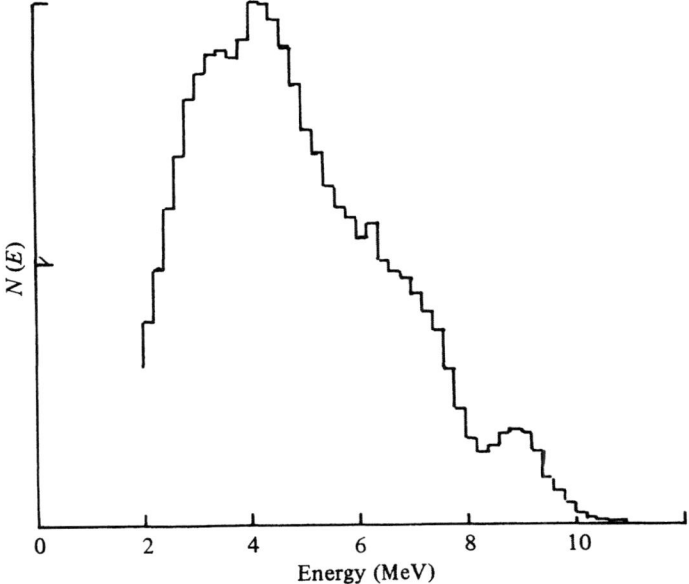

Such sources are quite inefficient at producing neutrons compared with the alpha-particle output. A one-curie (37 GBq) americium–beryllium source has a neutron yield of about 2.5×10^6 s^{-1}, whereas the alpha-particle production rate from the decay of ^{241}Am is 3.7×10^{10} s^{-1}.

Low energy accelerators can be a useful source of fast neutrons using beams of deuterons (^{2}H$^+$ ions) in the following two reactions.

(a) ^{3}H + ^{2}H $\rightarrow$ ^{4}He + n + 17.6 MeV

The neutron energy depends both on the angle of emission and on the deuteron energy but is around 14 MeV (14.1 MeV at 90° with little dependence on deuteron energy). A high yield is obtained with deuteron energies of only 100–150 keV.

(b) ^{2}H + ^{2}H $\rightarrow$ ^{3}He + n + 3.5 MeV

Again the neutron energy depends upon the deuteron energy and the angle of emission but is of the order of 2.5 MeV for low bombarding energies. Deuteron energies from about 150 keV up to a few MeV can be used for this reaction but the yield is much lower than for the first reaction above.

A very compact portable neutron source is the isotope ^{252}Cf. Apart from undergoing alpha-particle decay this isotope can also disintegrate by spontaneous fission, giving a broad spectrum of energies in the MeV range, as shown in Fig. 1.11. The main disadvantage of this source of neutrons is its short effective half-life of about 2.7 years. The specific neutron yield is high: one microgram of ^{252}Cf releases 2.3×10^6 neutrons per second.

Nuclear reactors are another source of neutrons having a very wide energy range. Fast neutrons with an energy spectrum, as shown in Fig. 1.11, are released by fission in the reactor fuel elements, but in all nuclear reactors except fast reactors a low mass number material, known as the moderator, such as water, heavy water or graphite, is deliberately incorporated so that the neutrons are slowed down by elastic collisions. This considerably increases the probability that a neutron can initiate fission in a further fissile nucleus (e.g. ^{235}U or ^{239}Pu) and so reduces the proportion of fissile material needed in the fuel compared with a fast reactor. Consequently such a nuclear reactor is a useful source of slow neutrons, which have a near Maxwellian velocity distribution that is characteristic of the temperature of the moderator. For a moderator at 20 °C (293 K) the most probable velocity is 2200 m s^{-1} which corresponds to an energy of 0.0253 eV. Hence a nuclear reactor can be a source of neutrons ranging in energy from almost zero to 10 MeV.

Fast neutrons will interact with matter mainly by elastic scattering, thus producing energetic ions, particularly with light scattering nuclei. The ionisation from the ion will therefore enable the neutron to be detected although the neutron cannot ionise directly since it has zero charge. If a neutron of energy E_0 and of relative mass 1 is elastically scattered from a nucleus of relative mass A (where to a close approximation A may be taken as the mass

Fig. 1.11. Neutron energy spectrum from a fission source.

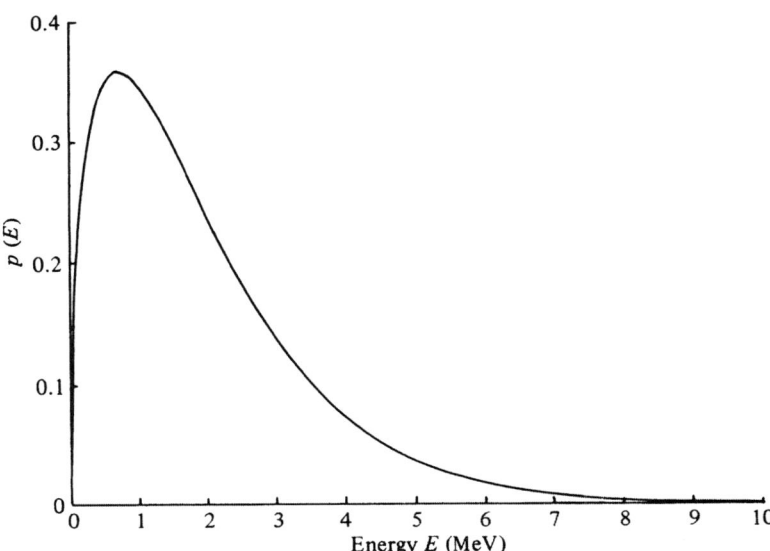

number) at an angle ϕ in the laboratory system then the energy of the scattered neutron can be shown to be

$$E(\phi) = E_0 \frac{A^2 + 1 - 2 \sin^2 \phi + 2 \cos \phi \sqrt{(A^2 - \sin^2 \phi)}}{(A + 1)^2}$$

(1.9)

The maximum energy loss possible in a collision with a nucleus $A > 1$ is for a scattering angle of 180°, and hence the maximum energy loss possible in a single collision is given by

$$\Delta E = E_0 - E(180°) = E_0 \frac{4A}{(A + 1)^2} \qquad (1.10)$$

For neutrons scattered from hydrogen nuclei (protons) for which $A = 1$ the maximum angle of scatter is only 90° at which the neutron has zero energy, and so for hydrogen equation (1.9) becomes:

$$E(\phi) = E_0 \frac{\cos^2 \phi + \cos \phi \, |\cos \phi|}{2} \qquad (1.11)$$

which gives $E(\phi) = 0$ for $90° \leq \phi \leq 180°$ and can be simplified to

$$E(\phi) = E_0 \cos^2 \phi \qquad (1.11a)$$

provided that ϕ is restricted to the range 0–90°. For neutron scattering from hydrogen nuclei all energies in the range 0–E_0 are equally probable for the scattered neutron. This is a particularly useful reaction for fast neutron detection since a fast proton is produced at an angle θ to the initial neutron direction and the proton energy as a function of this angle is given by

$$E_p(\theta) = E_0 \cos^2 \theta \qquad (1.12)$$

where θ is also restricted to the range 0–90°. Note that $\theta + \phi = 90°$ for all possible angles. Since the proton has a single positive charge it will cause ionisation, thus enabling the neutron to be detected, although the neutron will lose energy and change direction.

Slow neutrons have insufficient energy to produce protons capable of causing ionisation in an elastic scatter with a hydrogen nucleus. They can, however, take part with certain nuclei in nuclear reactions that release charged particles, thus enabling the neutron to be detected at the expense of its removal.

One such reaction is the fission reaction with, for example, ^{235}U, which releases two very energetic fission fragments totalling about 160 MeV in energy. Other common slow neutron reactions are

$$^6\text{Li} + \text{n} \rightarrow {}^3\text{H} + {}^4\text{He} + 4.8 \text{ MeV}$$
$$^{10}\text{B} + \text{n} \rightarrow {}^7\text{Li} + {}^4\text{He} + 2.8 \text{ MeV}$$
$$^1\text{H} + \text{n} \rightarrow {}^2\text{H} + \gamma + 2.25 \text{ MeV}$$

The first two reactions are important for slow neutron detection since they release energetic charged particles and the third is important for slow neutron interactions in living tissue since it results in a gamma-ray dose to the tissue.

Further reading
W. E. Burcham, *Nuclear Physics* (2nd edn), Longmans, 1973.

2

.. ..

Statistics of particle counting and dead-time

2.1 Statistics of particle counting

The rate of decay of a radioactive source can be characterised by the decay constant λ or by the half-life, which are related quantities. If a sample of a single species of radioactive isotope contains N radioactive atoms at a particular instant in time, and dN of these will decay in a small time interval dt, then N, dN and λ can be related by

$$dN = -N\lambda \, dt \tag{2.1}$$

which can be rearranged to give

$$dn/N = -\lambda \, dt \tag{2.1a}$$

This implies that in a small time interval dt, the fraction of atoms decaying is proportional to the time interval provided that $\lambda \, dt \ll 1$. Equation (2.1a) can be integrated to yield an expression for the number of radioactive atoms present at any time t after the start when N_0 atoms are present:

$$N = N_0 e^{-\lambda t} \tag{2.2}$$

A rather more useful concept than tha decay constant, which has the dimensions of reciprocal time, is the half-life. The half-life of a radioactive isotope is defined as the time taken for half of the initial atoms present to decay. This can be found by setting $N = N_0/2$ and $t = T_{1/2}$ in equation (2.2) to obtain

$$T_{1/2} = \ln 2/\lambda = 0.693/\lambda \tag{2.3}$$

After one half-life one-half of the initial number of active atoms remain, after two half-lives one-quarter remain and after three half-lives one-eighth remains, etc.

Although half-life or decay constant can be measured to a high precision, given the right experimental conditions, radioactive decay is really a statistical process. If a radioactive source is used which has a half-life sufficiently long that the fraction of atoms decaying over a period of a few hours is negligible, then the number of counts recorded on a detector in a sequence of equal time intervals will not be identical but will exhibit statistical fluctuations.

If p is the probability that one atom will decay in a finite time interval t (which from equation (2.2) is $1 - e^{-\lambda t}$) then the probability of n atoms out of a total N decaying is given by the binomial distribution:

$$P(n) = \frac{N!}{(N-n)!n!}\, p^n(1-p)^{N-n} \tag{2.4}$$

where n and N are whole numbers. Note that

$$\sum_{n=0}^{N} P(n) = 1$$

The arithmetic mean of the number decaying in a time t is given by

$$\bar{n} = \sum_{n=0}^{N} nP(n) = pN \tag{2.5}$$

It can also be shown that if the deviation of a number n from the mean $\bar{n}$ is defined by $\bar{n} - n$ then the root mean square deviation σ or the standard deviation is

$$\sigma^2 = \overline{(\bar{n}-n)^2} = Np(1-p) \tag{2.6}$$

This leads to the conclusion that if the time interval t is $\ll 1/\lambda$ then

$$1 - p \simeq 1$$

and the square of the standard deviation is then

$$\sigma^2 = Np = \bar{n} \tag{2.7}$$

Hence the standard deviation becomes the square root of the mean value

$$\sigma = \bar{n}^{1/2} \tag{2.8}$$

This is a very important conclusion and is widely used in the analysis of counting data, as will be described further in this chapter.

If the total number of active atoms is large and the probability of decay in the given time interval is $p \ll 1$ then the binomial distribution will approximate to the Poisson distribution, which also applies to integer numbers:

$$P(n) = \frac{\bar{n}^n e^{-\bar{n}}}{n!} \tag{2.9}$$

Just as for the binomial distribution the mean value $\bar{n} = pN$ and the root mean square deviation is given by $\sigma = \bar{n}^{1/2}$. Further simplifications can be applied if the mean value $\bar{n}$ is reasonably large and the probability distribution then approximates to the Gaussian distribution, which holds for both integral and non-integral values of n:

$$P(n) = \frac{1}{\sqrt{(2\pi\bar{n})}} \exp\left(-\frac{(\bar{n} - n)^2}{2\bar{n}}\right) \tag{2.10}$$

Now let the deviation $(\bar{n} - n)$ be denoted by ε and replace the mean value $\bar{n}$ by the mean square of the deviation σ^2. The Gaussian distribution can now be rewritten in terms of these two quantities:

$$P(\varepsilon) = \frac{1}{\sqrt{(2\pi)}} \cdot \frac{1}{\sigma} \exp\left(-\frac{\varepsilon^2}{2\sigma^2}\right) \tag{2.11}$$

This is the probability of having a deviation ε from the mean value, and ε, which may take on positive or negative values, is not restricted to integer values.

Table 2.1. *Integral of the Gaussian distribution*

$\pm Y$	$\int_{-Y}^{+Y} P(y)\, dy$
0	0
0.674	0.5
1.0	0.683
1.96	0.950
3.00	0.997

Although only the binomial distribution is exact for the analysis of counting measurements, the very close approximation provided by the Gaussian distribution is much more convenient for working out the consequences of statistical fluctuations in counting data. If the dimensionless quantity $\varepsilon/\sigma = y$ is taken as a new variable then the Gaussian distribution can again be rewrittten in the form:

$$P(\varepsilon)\ d\varepsilon = P(y)\ dy = \frac{1}{\sqrt{(2\pi)}} \exp\left(-y^2/2\right) dy \qquad (2.12)$$

and the integral of this function between symmetrical limits gives the probability of events being within a given deviation from the mean:

$$\int_{-Y}^{+Y} P(y)\ dy = \frac{1}{\sqrt{(2\pi)}} \int_{-Y}^{+Y} \exp\left(-y^2/2\right) dy \qquad (2.13)$$

In Table 2.1 some values of this integral are given and the form of the probability distribution is also shown in Fig. 2.1.

From the figures given in Table 2.1 it will be seen that 68.3% of the events lie within ± 1 standard deviation of the mean value and that almost all the events lie within ± 3 standard deviations of the mean.

Fig. 2.1. Gaussian probability distribution $P(y) = 1/\sqrt{(2\pi)} \exp\left(-y^2/2\right)$ where y is the deviation in units of standard deviations.

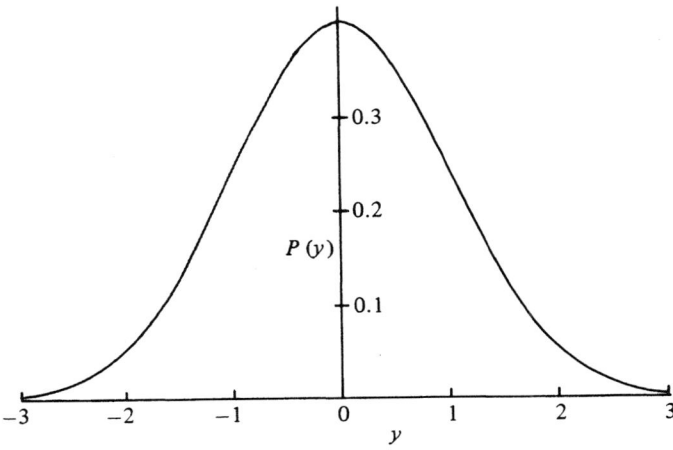

It is rarely possible to take a whole sequence of similar counts, either due to lack of time or due to decay of the source during the counting interval, so measurements generally have to rely on a single count. The probability distribution indicates that this single count has a 68.3% chance of being within ±1 standard deviation of the true mean and a 99.7% chance of being within ±3 standard deviations of the true mean. However, with a single count the true mean cannot be found but since the standard deviation is given by the square root of the true mean it can be closely approximated by taking the square root of the actual recorded count:

$$\sigma = \bar{n}^{1/2} \simeq n^{1/2}$$

Hence the uncertainty in a single count n is commonly given as $\pm n^{1/2}$, which is usually called the standard deviation. Less commonly the uncertainty is quoted as $\pm 3n^{1/2}$ and is regarded as being the extreme uncertainty.

It is often necessary to subtract the effect of background radiation from a measurement. Suppose that the measured total count, including background, is c and that the background count, recorded without any source present, is b. Firstly, it will be assumed that both these measurements are made over the same time interval t. It will also be assumed that there is negligible error in t, which is a valid assumption when automatic timing is used, a feature that is available on most counting modules nowadays. Each count will have a standard deviation:

$$\sigma_c = c^{1/2} \quad \sigma_b = b^{1/2}$$

The overall standard deviation is found by taking the square root of the sum of the squares of the individual standard deviations:

$$\sigma = (\sigma_c^2 + \sigma_b^2)^{1/2} \tag{2.14}$$

and the background corrected count n is therefore

$$n = c - b \pm (c + b)^{1/2} \tag{2.15}$$

If the count rate is needed the count is just divided by the counting time interval t but care must be taken in computing the standard deviation from count rates. Let $n' = n/t$, $b' = b/t$ and $c' = c/t$ be the count rates. The background corrected count rate is, from equation (2.15),

$$n' = \frac{(c - b)}{t} \pm \frac{(c + b)^{1/2}}{t} = c' - b' \pm \left(\frac{c' + b'}{t}\right)^{1/2}$$

(2.16)

If the total and background counts are recorded over different time intervals t_1 and t_2 respectively, where generally the background is recorded for a longer time interval than the source count, then the total count rate is $c' = c/t_1$ with a standard deviation of $c^{1/2}/t_1 = (c'/t_1)^{1/2}$. The background count rate is $b' = b/t_2$ with a standard deviation of $b^{1/2}/t_2 = (b'/t_2)^{1/2}$. The corrected count rate then becomes

$$n' = c' - b' \pm (c'/t_1 + b'/t_2)^{1/2}$$ (2.17)

In most cases the effect of background on standard deviation is small but when very weak sources are being counted the reliability of the results may be strongly affected by background. Consider a weak source that has a true count rate of only 10% of the background count rate and assume that source and background counts alone are determined for equal time intervals. Depending on the counting time the following results shown in Table 2.2 might be obtained.

In the first line of Table 2.2 the standard deviation of the corrected count is larger than the difference between total and background counts and so there can be no confidence that the difference count is even meaningful. Even for the longer counting times it will be observed that the uncertainty in the corrected count is large and that to reduce the uncertainty by a factor of ten requires a counting time 100 times as long.

Table 2.2. *Effects of counting time on standard deviation in the presence of background*

t	c	b	n	σ	$\sigma/n\%$
100	110	100	10	14.5	145
1000	1100	1000	100	45.8	45.8
10000	11000	10000	1000	144.9	14.5

Now consider a succession of counts taken over equal time intervals on a source having negligible decay over the total time taken for the experiment, and assume that four counts are taken:

$$n_1 \pm n_1^{1/2} \quad n_2 \pm n_2^{1/2} \quad n_3 \pm n_3^{1/2} \quad n_4 \pm n_4^{1/2}$$

The total count n is just the sum of the four individual counts

$$n = n_1 + n_2 + n_3 + n_4 \pm (n_1 + n_2 + n_3 + n_4)^{1/2}$$

and therefore the standard deviation is just $n^{1/2}$. No apparent statistical advantage is therefore obtained by splitting up a count into a series of shorter counts. In fact if manual timing is used there may be additional error introduced due to variations in each counting period since reaction time is variable and typically of the order of one-quarter to one-half of a second. The one possible advantage of using a sequence of counts (apart from demonstrating the presence of statistical fluctuations) is when intermittent interference is present. An abnormally high or low count as defined by $n > \bar{n} + 3\bar{n}^{1/2}$ or $n < \bar{n} - 3\bar{n}^{1/2}$ could then be rejected, whereas the spurious counts would not be identified in one long count.

2.2   Dead-time

A further source of error in recorded counts arises when the counting system has finite dead-time or paralysis-time. Dead-time can arise in several ways. The Geiger–Mueller counter (see Chapter 4) becomes dead for typically a few hundred microseconds following the recording of an event. It can also arise due to the finite time required to process pulses by discriminators, scalers and other electronic modules. In the multi-channel analyser the dead-time associated with the processing of a pulse is generally a function of pulse height (see Chapter 7). Allowance for this effect is usually made in the instrument by gating the timing clock so that it does not record time during a dead-time period. The indicated time is, therefore, the time for which the instrument is live, that is, the time for which it is able to record events.

For a system with a fixed dead-time, such as the Geiger–Mueller counter, a simple correction can be made to the observed count rate provided that the dead-time is known. Let c' be the recorded count rate (counts per second), r' be the true count rate that would be recorded in an ideal system with zero dead-time and let τ be the dead-time (seconds) where τ is typically of the order of 500 μs for a Geiger–Mueller counter.

During each second the system is dead for a time $c'\tau$ and hence in each second the system is live for a time $(1 - c'\tau)$ seconds. The fraction of the possible total number of randomly spaced counts recorded in each second is then $1/(1 - c'\tau)$ but this fraction is also c'/r'. Hence the true count rate is related to the observed count rate by

$$r' = c'/(1 - c'\tau) \tag{2.18}$$

Equation (2.18) can be rearranged to give the observed count rate in terms of the true count rate:

$$c' = r'/(1 + r'\tau) \tag{2.19}$$

From equation (2.19) the effect of dead-time can be seen more clearly. As the true count rate r' becomes very large the observed count rate c' tends towards an asymptotic value of $1/\tau$. The relation between c' and r' is shown in Fig. 2.2. It is obvious that if dead-time

Fig. 2.2. Recorded count rate as a function of input count rate for a fixed dead-time τ.

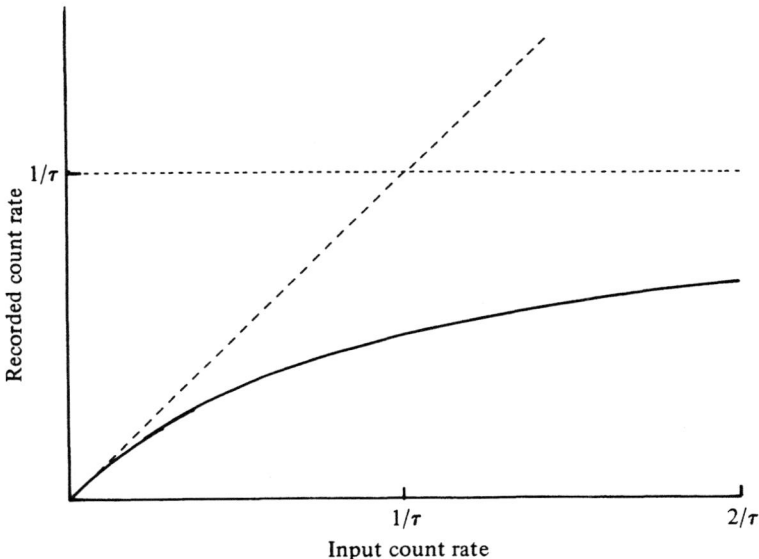

corrections are not made (except at count rates $\ll 1/\tau$) then serious underestimation of count rate is possible. An accurate estimate of dead-time is therefore essential. Some counting systems incorporate an electronically generated pre-set paralysis-time that is slightly greater than the counter dead-time, which may vary slightly from pulse to pulse. Measurement of dead-time on such a system can sometimes be carried out by injecting a suitable pulse train from a variable frequency signal generator into the input. The pulses from the signal generator are not randomly spaced in time and so equations (2.18) and (2.19) do not hold. For frequencies less than $1/\tau$ the measured count rate is exactly equal to the input frequency. As the signal generator frequency is gradually increased the observed counting rate will suddenly halve at a frequency of $1/\tau$, thus enabling the dead-time to be found.

When the foregoing procedure is not possible it is necessary to perform a sequence of counting experiments with the aid of two radioactive sources of approximately equal strength. This strength should be such that when a source is placed close to the counter the effect of dead-time is large. For example, two 20 μCi (740 kBq) sources of ^{60}Co are suitable when placed about 20 mm below the end window of an MX148 Geiger–Mueller counter. The sources must be accurately located with respect to the counter to avoid count rate changes due to positional errors. The general experimental arrangement is shown in Fig. 2.3. The sources are each counted separately in their appropriate locations and then counted

Fig. 2.3. Arrangement of Geiger–Mueller counter and sources for measurement of dead-time. The two sources *S*1 and *S*2 should be approximately equal in strength.

together for a time adequate to obtain good statistics to get count rates c_1', c_2', and finally c_3', for the combined sources. From equation (2.18) the true count rates are:

$$r_1' = \frac{c_1'}{1 - c_1'\tau} \qquad r_2' = \frac{c_2'}{1 - c_2'\tau} \qquad r_3' = r_1' + r_2' = \frac{c_3'}{1 - c_3'\tau}$$

Therefore, eliminating r_1', r_2' and r_3' from these equations,

$$\frac{c_1'}{1 - c_1'\tau} + \frac{c_2'}{1 - c_2'\tau} = \frac{c_3'}{1 - c_3'\tau}$$

from which the dead time can be obtained after rearrangement:

$$\tau = \frac{1}{c_3'}\left[1 - \sqrt{\left(1 - \frac{c_3'(c_1' + c_2' - c_3')}{c_1'c_2'}\right)}\right] \tag{2.20}$$

If the two sources were identical in strength and in geometrical positioning then the count rates would be equal, $c_1' = c_2' = c'$, and the expression for dead-time, equation (2.20), would simplify to

$$\tau = \frac{2c' - c_3'}{c'c_3'} \tag{2.21}$$

Provided that the two sources are similar in strength, although not exactly equal, then a close approximation for dead-time may be obtained by analogy with equation (2.21) and this turns out to be more accurate than expansion of equation (2.20):

$$\tau = \frac{2(c_1' + c_2' - c_3')}{(c_1' + c_2')c_3'} \tag{2.22}$$

The main term affecting the precision of measurement of dead-time in equation (2.22) is the numerator $(c_1' + c_2' - c_3')$. Assuming that all the individual counts have been taken over the same time interval t, then by analogy with equation (2.16) the uncertainty or standard deviation in τ is, to a first approximation,

$$\delta\tau = \frac{2\sqrt{\left(\dfrac{c_1' + c_2' + c_3'}{t}\right)}}{(c_1' + c_2')c_3'}$$

and so the fractional uncertainty in dead-time is approximately

$$\frac{\delta\tau}{\tau} = \frac{\sqrt{\left(\dfrac{c_1' + c_2' + c_3'}{t}\right)}}{(c_1' + c_2' + c_3')} \tag{2.23}$$

For counting systems using Geiger–Mueller counters the preceding treatment of dead-time and its correction, using a fixed or non-extendable dead-time is applicable. Most other detector types do not possess an inherent dead-time, although pulse pile-up and associated electronic processing and recording modules can cause count losses at very high count rates. Certain specialist amplifiers that enable very accurate measurement of pulse size to be made have an electronically generated dead-time that is extended if a further pulse is received during the current dead-time period. The purpose of this is to prevent pulse pile-up, which would distort the shape of the original pulse and so give a false measurement of pulse height. This extension of dead-time is not limited to just one following pulse but will reject a sequence of pulses arriving with time spacings less than the minimum dead-time. At high count rates the losses caused by such a system will be significantly higher than for a system with a fixed dead-time and equation (2.16) cannot be used for dead-time correction.

From the Poisson distribution (equation (2.9)) it is possible to find the probability that no events will occur in a time T in which an average of $r'T$ counts would be expected, where r' is the average count rate:

$$P(0) = \frac{e^{-r'T}}{0!} = e^{-r'T}$$

The probability of an event occurring in a time dT following this is therefore

$$P(0) \cdot r' \, dT = P_1(T) \, dT$$

and therefore

$$P_1(T) \, dT = r' e^{-r'T} \, dT$$

This is the distribution function for intervals between adjacent random counts with an average rate of arrival r' and it is just a simple exponential function since r' is a constant. The probability that the interval will be greater than a time τ is given by

$$P(\tau) = \int_{\tau}^{\infty} P_1(T) \, dT = e^{-r'\tau}$$

Since with extendable dead-time only pulses following the previous pulse, whether recorded or not, at a time interval greater than the minimum dead-time τ can be recorded, the recorded count rate becomes $c' = $ (probability of interval $> \tau$) $\times$ (arrival rate), giving

$$c' = r'e^{-r'\tau} \tag{2.24}$$

This is an awkward correction to apply since the expression cannot be solved explicitly for r' when c' is known since r' occurs both in the exponent and as a multiplier of the exponential. Fig. 2.4 shows the relation between c' and r' for extendable dead-time and should be compared with Fig. 2.2 for non-extendable dead-time. It will be seen that the observed count rate c' has a maximum value of $0.368/\tau$ for an input count rate of $1/\tau$, and that for observed count rates less than the maximum there are two possible values of true count rate r'. It is up to the experimenter to ensure that he knows which part of the correction curve is the one that applies to his experimental count rate.

The transcendental equation for r', as given in equation (2.24), can be easily solved for the case where $r' < 1/\tau$ by recasting in the form

$$r' = c'e^{r'\tau} \tag{2.25}$$

Fig. 2.4. Recorded count rate as a function of input count rate for an extendable dead-time with minimum value τ.

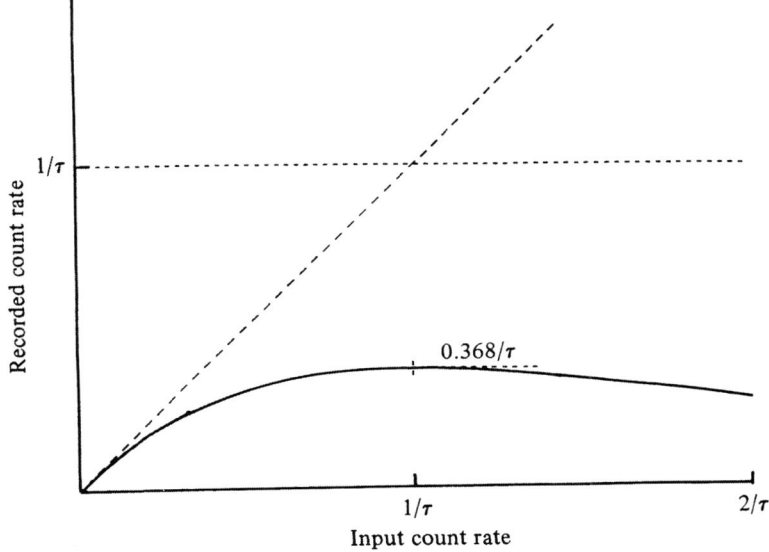

A guessed value of r' ($<1/\tau$) is inserted in the exponent and a new value of r' is found. This new value is inserted into the exponent and the process is repeated. After a few iterations r' settles down to a constant value.

If r' is greater than $1/\tau$ then the iterative process using equation (2.25) diverges. A convergent solution can be obtained in this case by recasting equation (2.24) in the form

$$r' = (1/\tau) \ln (r'/c') \tag{2.26}$$

and following the same sort of iterative procedure.

2.3   Fano factor

The previous sections of this chapter have been concerned with the statistics of counting events. In assessing the performance of some detectors it is necessary to know the statistical fluctuations in the size of a nominally constant amplitude pulse in order to determine the energy resolution. Since the pulse is produced by a finite number of charge carriers it might be expected that the standard deviation would just be equal to the square root of the mean number of charge carriers. Particularly in semiconductor and gas-filled detectors this is not true since the average energy needed to produce an ionisation is only a few times the threshold energy for ionisation, and consequently the standard deviation is reduced. Calculation of the magnitude of this effect is difficult but the conclusions can be expressed as

$$\text{standard deviation} = (FN)^{1/2} \tag{2.27}$$

where N is the average number of charge carriers released and F is the *Fano factor*, a number less than unity. For semiconductor detectors the Fano factor F is of the order of 0.1 and so the standard deviation is reduced by a factor of about 3 compared with the simplistic view. This reduction in standard deviation has an important effect on the energy resolution of such detectors, although, as will be seen in Chapter 6, there are other effects that somewhat increase the standard deviation.

Further reading

J. R. Taylor, *An Introduction to Error Analysis*, OUP, 1982.

3

... .

Gas-filled detectors

3.1 Introduction

Gas-filled detectors utilise the ionisation produced in a gas by the interaction of ionising radiation in order to produce an electrical signal that indicates the presence of radiation and, under suitable circumstances, its energy and type. Although different gases have different first ionisation potentials the average energy expended per ionisation in most commonly used gases is remarkably constant at about 35 eV. Table 3.1 gives the first ionisation potential of some of the common gases. The amount of energy released in an individual ionising event is very variable and occasional electrons of several keV energy are found (these are known as delta-rays). Consequently there will be statistical fluctuations in the total number of ionisations caused by a succession of equal energy particles stopped in the gas. If the average number of

Table 3.1. *Ionisation potential of common gases*

Gas	Ionisation potential (eV)
Hydrogen	13.56
Helium	24.59
Nitrogen	14.53
Oxygen	13.62
Neon	21.56
Argon	15.75

ionising events is N then the standard deviation will be close to $N^{1/2}$. This is for primary ionisations and is different from the effects observed in solids (for example, semi-conductor detectors, see Chapter 6) where the energy spread is much smaller. A similar effect is observed in gas-filled detectors in which multiplication of the primary charge occurs and then the standard deviation can be expressed as $(FN)^{1/2}$ where F is the Fano factor (see Section 2.3) and N is the total number of charges collected. For gas-filled detectors with multiplication the Fano factor is in the range 0.1 to 0.2. The typical amount of primary charge released can be calculated using an average figure of 35 eV expended per ionisation. Consider an alpha-particle of energy 5.25 MeV. The average number of ion-pairs produced when the particle loses all its energy is 1.5×10^5 and hence the charge of either sign is $1.5 \times 10^5 \times 1.6 \times 10^{-19}$ C $= 2.4 \times 10^{-14}$ C (or 4.57×10^{-15} C per MeV). This small quantity of charge can be difficult to measure accurately. Two types of detector are possible: mean current detectors and pulse detectors and these will be considered in more detail.

3.2   Mean current detectors

Large volume detectors filled with dry air or nitrogen at atmospheric pressure are sometimes used for continuous measurement of gamma-radiation levels in working areas. Gamma-ray dose rate levels up to 2.5 mrad h^{-1} (25 μSv h^{-1}) can be expected under normal working conditions (see Chapter 8). With a detector of sensitive volume 10 litres, this normal upper level would produce a current of 2.5×10^{-12} A if suitably polarised electrodes are included in the volume, and since lower dose rate levels must be detected an ability to measure down to 10^{-13} A is necessary. Such current levels are measurable but great care must be taken with insulators to avoid surface leakage currents by such means as a guard ring. The electrodes must have sufficient voltage difference between them so that the electric field within the detector is high enough to prevent ions and electrons from recombining.

The other major application of mean current detectors is in

nuclear reactor instrumentation. If the detector is lined with a boron compound then the reaction

$$^{10}\text{B} + \text{n} \rightarrow {}^7\text{Li} + {}^4\text{He} + 2.8 \text{ MeV}$$

can occur, principally with slow neutrons. This reaction will then release either the ^7Li ion or the ^4He ion (alpha-particle) into the gas filling (not both ions since they travel in opposite directions), and so will produce a mean current that is proportional to the neutron flux which is proportional to reactor power. Currents can be relatively large at full power but unfortunately a simple boron lined ionisation chamber will also respond to the intense gamma-radiation present in a nuclear reactor. This would not matter if the intensity of gamma-radiation were proportional to reactor power but it mostly depends upon the running history of the reactor, thus leading to possibly very misleading readings.

One solution to this problem is the compensated ionisation chamber, which is a combination in one casing of two mean current ionisation chambers. One of these is neutron sensitive and has a boron lining but the other does not contain boron. It is designed, therefore, to be insensitive to neutrons but to have the same response to gamma-radiation. The two chambers are electrically connected in such a way that only the current due to neutron interactions is recorded, the current due to gamma-radiation in each chamber being mutually cancelling. Fig. 3.1 shows a simplified circuit for such a compensated ionisation chamber together with the typical construction of such a device.

Ionisations caused by the charged particles from the slow neutron reaction with boron-10 will cause a current to be recorded by the electrometer. If equal amounts of ionisation are produced by gamma-radiation in each section of the detector then positive ions in the $(n + \gamma)$ section will be collected at the central electrode and an equal current of electrons will arrive at the central electrode from the $(\gamma \text{ only})$ section. Thus there is no net current at the central electrode and no current will flow to the electrometer. Current flow due to gamma-radiation is only via the outside electrodes and this current finds its return path through the polarising supplies.

3.3   **Pulse detectors**

3.3.1   *Ionisation counters*

Most radiation detectors, gas-filled or otherwise, are pulse detectors. That is, they detect the interaction of individual alpha-particles, beta-particles, gamma-rays, neutrons, etc. Returning to the example of a 5.25 MeV alpha-particle the total charge available for collection is 2.4×10^{-14} C, as has already been calculated. Two types of charge carrier exist in the gas filling; positive ions which are heavy and move slowly and electrons which move much more rapidly. Since both positive ions and electrons have short collision mean free paths in the gas (ions $\sim 5 \times 10^{-4}$ mm, electrons $\sim 2 \times 10^{-3}$ mm at atmospheric pressure) then the velocity gained by acceleration in the electric field over one mean free path is small compared with the random thermal velocity. This small increase over the random thermal velocity will be lost at the next collision and so the charge carriers will drift in the direction of the electric

Fig. 3.1. Compensated ionisation chamber for detection of slow neutrons in the presence of gamma-radiation.

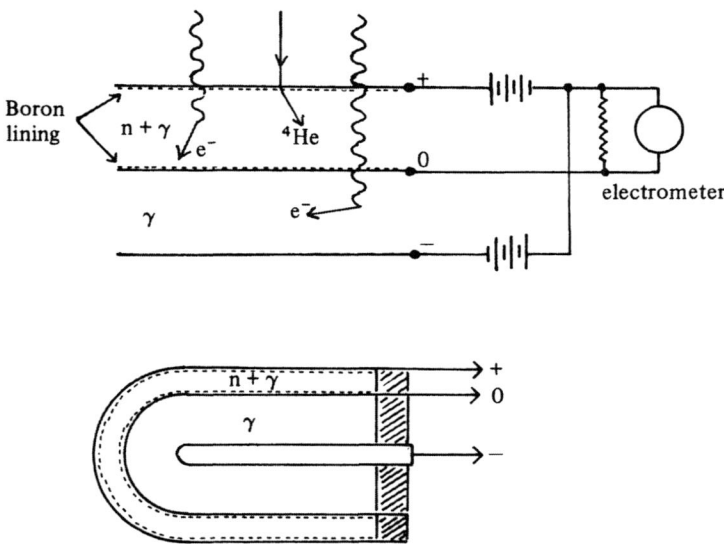

field with a drift velocity that is proportional to the electric field. The constant of proportionality between the drift velocity and the electric field is known as the mobility. Electrons have a higher mobility than positive ions and so the electrons will move rapidly to the positive electrode or anode and the positive ions will move much more slowly to the negative electrode or cathode. The current will therefore consist of a fast and a slow component and the overall current waveform will depend upon the position relative to the electrodes where the ionising events take place. To illustrate this point consider a plane–parallel pair of electrodes and a collimated beam of alpha-particles travelling parallel to the plates. Fig. 3.2 shows the situation for the alpha-particle beam close to the cathode and also close to the anode. As the system is effectively a parallel plate capacitor the moving charges will cause a current to

Fig. 3.2. Charge collection and pulse shape in a parallel plate ionisation chamber.

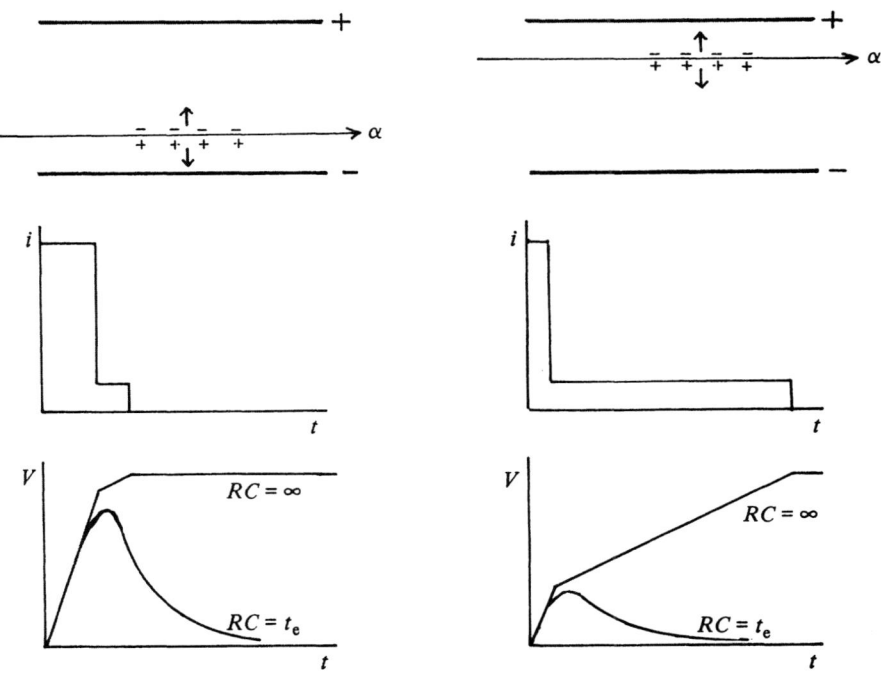

flow in the external circuit, electrons giving a high current for a short time and positive ions a low current for a long time. The time for which the current flows is proportional to the distance that the charge carriers have to travel before collection and so the current waveform is quite different for the two cases considered. The total charge collected is, however, the same in both cases and is equal to the total charge of either electrons or ions and not the sum of the two. Since the total charge is small and the capacity of the detector and connecting cables is finite the size of voltage pulse developed will be very small and considerable external amplification of the pulse will be necessary in order to record it. If the amplifier has an infinite time constant ($RC = \infty$) then the final voltage reached will only depend on the total charge and not on the position of the ionisations, but the time taken to reach the final voltage will depend upon position. For repetitive counting of events it is necessary to allow the amplified voltage pulse to decay away and so amplifiers with finite time constants must be used. If fast counting is required then the time constant must be similar in magnitude to the electron collection time. As a result only the electron portion of the pulse is effectively recorded and the pulse size varies with particle position. This effect is also indicated in Fig. 3.2 for a time constant equal to the electron transit time across the cathode to anode distance ($RC = t_e$). Such variation in pulse size for a constant amount of charge produced by ionisation is unfortunate since any information about the particle energy, which is proportional to the total ionisation, is completely lost.

The gridded ionisation chamber is a method of overcoming this disadvantage of parallel plate ionisation chambers and the general features are shown in Fig. 3.3. This device is only of use if the ionising particles can be restricted to the space between the grid and the cathode. If the grid is made sufficiently fine then nearly all the electrons will pass through the grid without being collected and will all travel the same distance from the grid to the anode. Consequently if the amplifier input is arranged between grid and anode an electron pulse that is proportional to the particle energy will be recorded.

In the early days of nuclear counting the ionisation chamber proved very difficult to use because of the problems of making high gain and low noise electronic amplifiers with thermionic valves mainly designed for radio use. Later developments, particularly the introduction of solid state amplifiers has made the use of ionisation chambers much more feasible, but for many applications solid state detectors (Chapter 6) are much more satisfactory.

The fission counter, which has a thin internal coating of fissile material such as ^{235}U or ^{239}Pu, is normally operated as an ionisation chamber, but since the average energy input to the gas filling from a fission fragment is 80 MeV the signal is much larger than that from other ionising radiations such as alpha-particles. Very good discrimination against gamma-radiation is obtained because of the very great difference in pulse size between neutron induced and gamma-ray induced events.

Consider again the 5.25 MeV alpha-particle stopped in an ionisation chamber and assume that the detector and cable connecting it to the input of a voltage sensitive amplifier has a capacity of 100 pF. The maximum voltage that could be produced at the amplifier input by detection of the alpha-particle, even with an infinite time constant, would then be 2.4×10^{-14} C/10^{-10} F = 0.24 mV. This magnitude of signal proved difficult to amplify above the inherent electrical noise of early valve amplifiers, so some method was sought of increasing the collected charge to improve the signal to noise ratio. As a result the proportional counter was developed.

Fig. 3.3. The gridded ionisation chamber.

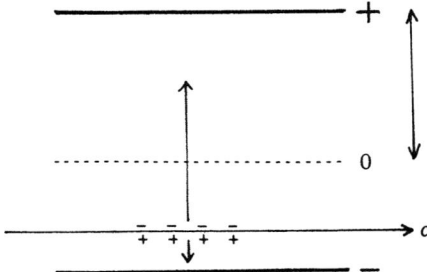

3.3.2 *Proportional counters*

If a very high electric field can be produced in the gas-filled detector, such that electrons can gain enough energy in one mean free path to cause ionisation at the next collision, then charge multiplication is possible. For a detector filled with argon at about one-quarter of atmospheric pressure the electron would need to gain about 20 eV in a distance of around 2×10^{-3} mm, thus requiring an electric field of about 10^7 V m^{-1}. This could not be achieved in a parallel plate detector of finite dimensions due to both the excessively high voltage required and an electric field that would exceed the breakdown field of the gas. In reality, ionisation by collision will become significant at somewhat lower field strengths than deduced above due to the statistical fluctuations in the actual collision path length, but this does not alter the conclusions about parallel plate detectors.

A coaxial detector arrangement having a fine wire anode stretched along the axis of a cylindrical cathode can generate high electric fields in the region close to the anode for only moderate polarising voltages. For such a system which has an anode of radius a and a cathode of radius c with a voltage between anode and cathode of V_a the voltage at a distance r from the centre is, in the absence of space charge,

Fig. 3.4. Electron multiplication by collision in the high electric field close to the anode surface.

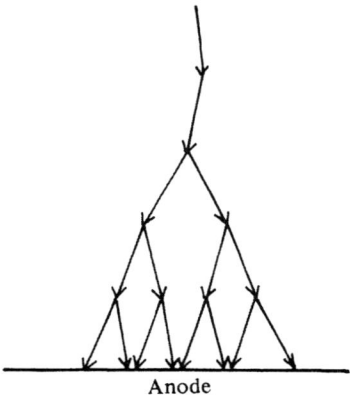

Anode

$$V_r = V_a \ln (c/r)/\ln (c/a) \tag{3.1}$$

and the electric field at this radius is

$$E_r = -dV/dr = V_a/(r \ln (c/a)) \tag{3.2}$$

Hence if the anode radius is taken to be $a = 2 \times 10^{-2}$ mm and the cathode radius is taken to be $c = 10$ mm then from equation (3.2) the anode voltage V_a will only need to be about 1250 V to give the required field of 10^7 V m^{-1} close to the surface of the anode and the field will fall to 10^7 V m^{-1} at a distance of 5×10^{-3} mm from the anode surface (2.5 mean free paths). Electron multiplication by collision is now clearly possible in the region close to the anode wire for easily produced polarising voltages.

If the multiplication takes place over an average of n mean free paths from the anode then the electron multiplication by collision is 2^n. Fig. 3.4 shows the multiplication process for $n = 3$. In practice, multiplication factors up to 100 are easily obtained and the output pulse is proportional in size to the initial ionisation. This gives rise to the name proportional counter for such a detector. Multiplication will vary with the applied anode voltage and Fig. 3.5 indicates the manner in which the pulse height varies with anode

Fig. 3.5. Pulse height versus anode voltage for a gas filled detector in the ionisation chamber and proportional regions.

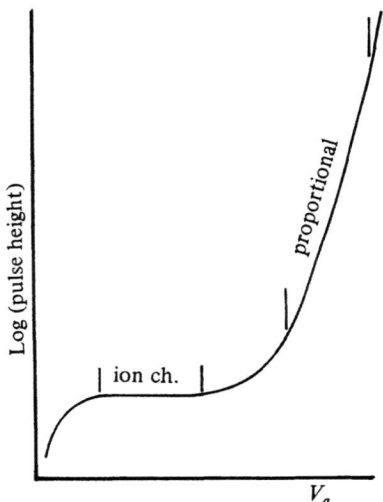

voltage from very low voltages at which recombination can occur, through the ionisation chamber region to the proportional region. Since the electrons, except the primary ones, are generated close to the anode they will only move through a small potential difference before being collected and so will only make a minor contribution to the voltage pulse, the major part of which is developed as the positive ions move towards the cathode. The pulse length is comparatively long and is determined by the transit time of the positive ions from the anode to the cathode:

$$\text{ion velocity} = \frac{\mu E}{p} = \frac{dr}{dt} = \frac{\mu V_a}{pr \ln (c/a)} \tag{3.3}$$

where μ is the ion velocity for unit field and unit pressure (the mobility of the ion) and p is the gas pressure. Hence the ion transit time T can be found from equation (3.3) by rearrangement and integration:

$$T = \frac{p \ln (c/a)}{\mu V_a} \int_a^c r \, dr = \frac{p \ln (c/a)}{2\mu V_a} (c^2 - a^2) \tag{3.4}$$

Since the majority of the pulse is developed in the early stages of the ion motion, where the voltage is changing rapidly with radius, then a short amplifier time constant can be used compared with the typical 100 μs collection time for the ions. Fig. 3.6 shows the voltage pulse as a function of time for both an infinite amplifier time constant and for a more realistic time constant of one-fifth of the ion collection time. From this it will be seen that for a detector of capacity C farads with N primary electrons of charge e coulombs and a multiplication factor M, the pulse height is approximately

0.5 MNe/C volts

One factor in electron multiplication has so far been neglected. Some of the ions produced in the secondary collision processes may be produced in an excited state. These excited ions can then decay to the ground state by emission of a quantum of radiation which can be in the ultra-violet region and which may therefore have enough energy to cause further ionisation by photoelectric absorption either from the cathode or the gas. Let the probability that a positive ion will produce a further electron by photoionisation be ε

which is generally $\ll 1$. For each primary electron there will be M secondary electrons and M ions and so there will be εM tertiary electrons. These will in their turn be attracted to the anode and multiplied by collision in the same way as the primary electrons, producing a further εM^2 electrons and positive ions that will lead to $\varepsilon^2 M^2$ further electrons still, and so on. The total number of electrons collected per primary electron will therefore be

$$M + \varepsilon M^2 + \varepsilon^2 M^3 + \varepsilon^3 M^4 \ldots = M/(1 - \varepsilon M) \qquad (3.5)$$

which is the true multiplication factor. For most proportional counters the true multiplication will differ very little from M but the effect becomes more significant as the voltage is increased beyond the range where true proportional behaviour holds.

Despite the extensive development of detectors since the invention of the proportional counter, a few examples are still in general use. The most common of these is the slow neutron detector known as the borontrifluoride counter or BF_3 counter from its gas filling.

Fig. 3.6. Voltage pulse as a function of time for a proportional counter showing the effect of amplifier time constant.

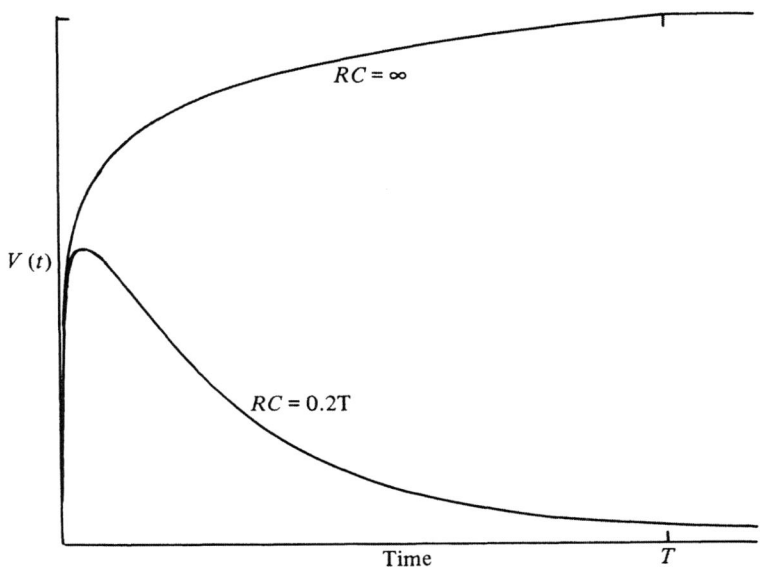

Borontrifluoride, commonly containing the separated isotope ^{10}B, instead of natural boron that only contains 22% of ^{10}B, is the sole filling gas and provides the neutron detection mechanism via the slow neutron reaction

$$^{10}\text{B} + \text{n} \rightarrow {}^{7}\text{Li} + {}^{4}\text{He} + 2.8 \text{ MeV}$$

Since the reaction takes place in the filling gas both the emitted particles contribute to the ionisation. Detector sizes range from about 6 mm diameter by 50 mm long up to 50 mm diameter by 1000 mm long. This large sized detector is used to monitor for large scale leaks in the fuel element cans of Magnox power reactors by detecting the delayed neutrons emitted by certain of the fission products that would be carried in the coolant gas stream under fault conditions. As the delayed neutrons have energies around 0.5 MeV the counters have to be enclosed in a polythene block to slow the fast neutrons down by elastic collisions with the hydrogen in order to obtain efficient detection.

At a filling pressure of about half an atmosphere the charged particles released in the nuclear reaction will be stopped in the gas. Gamma-ray interactions produce fast electrons mainly from the copper walls of the detector and these will only lose a small amount of energy in crossing the gas. Hence gamma-ray induced pulses will be much smaller in amplitude than neutron induced pulses and can be rejected by discriminating against pulses less in height than the neutron events. Slow neutrons can therefore be detected in the presence of high gamma-ray backgrounds and the limit is set by multiple pulse pile-up of gamma-ray events that give sum pulses of the same size as neutron events.

An alternative neutron detector is the ^{3}He proportional counter. This has a filling of ^{3}He only and detects slow neutrons by the reaction

$$^{3}\text{He} + \text{n} \rightarrow \text{p} + {}^{3}\text{H} + 0.765 \text{ MeV}$$

which has a reaction cross-section somewhat higher than the slow neutron reaction with ^{10}B. Since the stopping power of ^{3}He is lower than for BF_3 at the same pressure then the range of the reaction products may not be small compared with the dimensions of the counter and so loss of part of the particles' range in the walls may affect the pulse height spectrum of the device. This can be overcome to some extent by the use of higher filling pressures provided that they are not too high to prevent proportional multiplication. Due to the lower energy release of the reaction discrimination against gamma-radiation is not as good as for the BF_3 counter.

Both the BF_3 counter and the ^3He counter are most suited for detection of slow neutrons since in this energy range their reaction cross-sections are highest. ^3He detectors can, however, be used with lower efficiency for fast neutron detection up to about 1 MeV. The upper energy limit is set by elastic scattering of ^3He with neutrons: in a head on collision the ^3He nucleus will recoil with 0.765 MeV of energy from a neutron of 1.02 MeV. Hence neutrons of this energy and above give rise to pulses that will be confused with the detection of slow neutrons.

Elastic scattering from hydrogen is a widely used reaction for detecting fast neutrons. A proportional detector filled with hydrogen will respond to fast neutrons by producing fast knock-on protons. For mono-energetic neutrons the proton energies will be distributed uniformly over the range zero to the incident neutron energy (see equation (1.12)). The low stopping power of hydrogen for protons is, however, a disadvantage since the detector needs to have large dimensions. Methane can be used as an alternative filling gas to provide a higher stopping power and so a more compact detector, but the pulse height spectrum will then include a lower energy contribution from recoiling carbon nuclei, which can make determination of the neutron spectrum difficult except for mono-energetic neutrons. One possible method of determining the neutron spectrum from the pulse height spectrum is by differentiation. Because of the integral nature of the recoil, proton energy spectrum information can be lost about lower energy neutrons if normal pulse height discrimination is used to reject gamma-ray events. Another possibility for discrimination against gamma-radiation exists in this type of detector. Since the rise time of the signal depends on the radial distribution of the primary ionisation, then, provided that the proton range is small compared with the radius of the counter, proton pulses will have a faster rise time than electron induced pulses from gamma-ray interactions since the electron will generally have a path length greater than the counter dimensions. Pulse shape discrimination can then be used to separate neutron and gamma-ray induced events and the methods used for pulse shape discrimination with scintillation counters, as discussed in Chapter 5, are also applicable here.

Low energy X-rays, such as the characteristic X-rays from copper, cobalt or molybdenum targets used in X-ray crystallography, may be detected with proportional counters filled with a high atomic number gas such as argon, krypton or xenon. A thin beryllium or mica entrance window is needed for the X-rays in order to minimise losses by absorption before entering the active part of the detector. The filling gas should be chosen to have its K absorption edge (equal to the ionisation potential of the K-shell) greater than the X-ray energy to be detected. If this is not so then a characteristic X-ray will be generated by the photoelectric process which may escape from the detector without interacting. Since the electron released in this photoelectric interaction will also have a low energy, equal to the X-ray energy minus the ionisation potential of the K-shell, then the efficiency of the detector will be greatly reduced (K edges for Ar, Kr and Xe are at 2.97 keV, 12.6 keV and 29.7 keV respectively). This type of detector may, particularly in X-ray spectrometers, be more attractive than semiconductor detectors (see Chapter 6) since they are compact and do not need cooling to liquid nitrogen temperatures.

4

··· ··· ··· ··· ··· ··· ··· ··· ··· ··· ··· ··· ··· ··· ··· ··· ··· ··· ···

The Geiger–Mueller counter

4.1   Multiplication processes

The Geiger–Mueller counter (commonly shortened to the Geiger counter) is closely related to the proportional counter (see Section 3.3.2). If the voltage applied to a proportional counter is increased above the normal operating range, then, firstly, a region of higher multiplication is reached where the output pulse is no longer proportional to the primary ionisation (the region of limited proportionality). As the voltage is increased further, then eventually very large output pulses are produced which are of constant size and not related to the amount of primary ionisation. It has already been shown in Chapter 3 that the multiplication factor of a proportional counter is given by

$$\text{multiplication factor} = M/(1 - \varepsilon M) \qquad (3.5)$$

where M is the multiplication factor by electron collision and ε is the probability that an ion produced in a collision process will produce a further electron by photoionisation. Since M increases with applied voltage it is eventually possible for the product εM to exceed unity and so the multiplication factor (equation (3.5)) will initially tend to infinity, resulting in a runaway avalanche of ionisation. This runaway will not continue indefinitely since there will be some limiting process associated with it. Consider for the present that the counter anode is a fine wire stretched along the axis of a cylindrical cathode. All the multiplication will take place in the region closely surrounding the anode wire. The electrons are much lighter than the ions of the filling gas and since they only have a short distance to travel to the anode they will be collected in a time very short compared with the transit time of the positive ions from the anode region to the cathode, a time which is typically a few hundred microseconds. Consequently, a dense cloud of positive ions rapidly builds up around the anode. In a short time this cloud of positive ions, or space charge, will become sufficiently dense to affect the electric field around the anode, lowering the field below the value needed for runaway multiplication and thus terminating the multiplication process. The counter now becomes insensitive, or dead, until the positive ions have been cleared away. Just as in the proportional counter the output pulse is developed as the positive ions move towards the cathode. The multiplication is so large that pulse sizes are sufficient to operate simple scalers or ratemeters without the need for external amplification. This is the great advantage of the Geiger–Mueller counter since it enables simple light portable counting equipment to be made. Its main disadvantages are, firstly, the absence of any information about particle energy or type from the pulse size since pulse size is limited by space charge build up and not primary ionisation and, secondly, large dead-time effects at higher counting rates. If the voltage of the counter is increased there is little change in the counter characteristics until the voltage becomes sufficient for internal breakdown to occur, at which point counting rates become meaningless and the counter may be severely damaged. Consider the simple experimental arrangement shown in Fig. 4.1. The scaler

Fig. 4.1. Simple counting arrangement for a Geiger–Mueller counter.

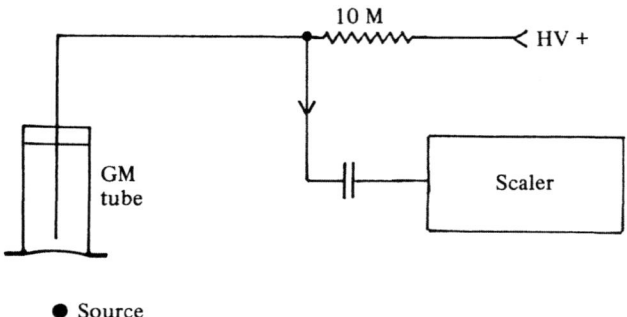

requires a minimum size of voltage pulse to trigger it, typically of the order of one volt, so for lower anode voltages at which the counter will operate in the proportional region no counts will be recorded since the pulse size is too small to operate the scaler. As the anode voltage is increased a point will be reached where avalanche multiplication starts and the output pulses become large enough to operate the scaler. This starting voltage is known as the threshold voltage, and as the anode voltage is increased above this value a counting rate versus voltage characteristic of the form shown in Fig. 4.2 is obtained. The voltage is termed HV in the figures since potentials in the range 300–2000 V may be required according to counter design. In Fig. 4.2 a region of almost constant count rate against voltage, called the plateau, will be observed. This plateau is not flat since the effective sensitive volume of the counter increases with voltage as a result of non-uniform axial electric fields at the ends of the anode wire. These end fields extend the length over which the avalanche field exists as the voltage increases. If an operating point is chosen near the centre of the plateau then a highly stabilised high voltage supply is not needed. Small drifts in high voltage will then have little effect on count rate. This is in strong contrast with the proportional counter, which needs a highly stabilised high voltage supply if pulse height is

Fig. 4.2. Voltage and count rate characteristics for a Geiger–Mueller counter.

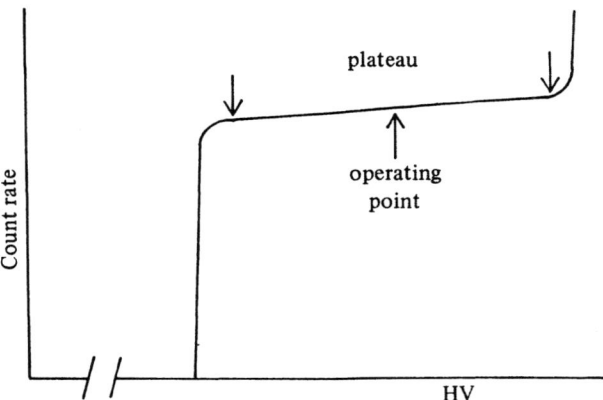

required to provide information about the ionising radiation. Since the Geiger–Mueller counter only needs a simple high voltage supply this is also an advantage for manufacturing simple portable equipment.

4.2   Multiple pulsing

A very large number of positive ions drift towards the cathode under the influence of the electric field. These will also have the random thermal velocity distribution as well as the comparatively low drift velocity and, consequently, a small proportion of these ions will have sufficient energy to eject electrons from the cathode on impact. Since the positive space charge has at this time been removed these electrons can be multiplied just as if they originated from the interaction of ionising radiations, and as a consequence multiple or spurious pulsing can occur. This multiple pulsing must be suppressed or quenched if meaningful measurements are to be made. Two methods of quenching may be used and these are frequently used together. One is internal quenching, achieved by careful choice of the filling gas mixture, and the other is achieved by an external electronic circuit.

4.2.1   *Electronic quenching*

By reference to Fig. 4.2 it can be seen that if the anode voltage is dropped below the threshold for a time somewhat longer than the ion collection time, then the electrons arising from ion impact cannot cause an avalanche and so no detectable pulse will occur. A separate circuit is used to drop the anode voltage below the threshold for a preset time and this is triggered by the leading edge of the output pulse. Such a feature is generally incorporated in non-portable equipment and has the advantage of providing an exactly constant dead-time per pulse since it is fixed electronically. Ion collection time is somewhat variable and so a more reliable dead-time correction can be made with an electronic quench. See Section 2.2 for dead-time correction and for methods of measuring the dead-time of a system. A further advantage of electronic quenching is that the pulse to operate the scaler can be derived from the quench circuit and so can be better matched to the scaler input requirements.

4.2.2 Internal quenching

Most commercially available Geiger–Mueller counters have an addition to the main filling gas in order to quench secondary pulsing. Since the main filling gas is generally monatomic, either argon or an argon–neon mixture, a quenching agent, which is a polyatomic gas or vapour with a lower ionisation potential than the main filling gas, is added. This quenching agent works in the following way: because of the relative proportions of main gas and quenching agent the ionisation in the avalanche will be predominantly in the main gas and so will consist of monatomic ions. During the time that the ions are drifting towards the cathode they will make many encounters with other gas atoms and molecules and charge exchange will take place with the quenching molecules since they are chosen to have a lower ionisation potential. By the time the ions reach the cathode they will almost all be ionised quenching molecules. Some of these ions will be energetic but instead of ejecting electrons from the cathode on impact they will tend to break up, and so multiple pulsing is suppressed. Quenching agents such as ethyl alcohol vapour, ethyl formate vapour, or halogens may be used.

4.3   Counter construction

Originally, organic vapours such as ethyl alcohol or ethyl formate were used as the quenching agent. These have the disadvantage that they are gradually consumed by the counting process and so, eventually, the counter fails. Most commercially available counters nowadays use small amounts or Cl_2 and Br_2 as the quenching agent. A halogen quenching agent has the advantage that the broken molecules will recombine and so counter life is much longer than for organic quenched counters. Halogens are, however, chemically very reactive so the materials used to construct the counter must be carefully chosen to be non-reactive. Stainless steel is normally used for the cathode, mica for end entrance windows, tungsten for the anode and glass for insulators. There is also a commonly used type of counter for counting liquid samples. The counter envelope is made of thin glass with a tungsten spiral as the cathode. This enables beta-radiation from the surrounding liquid sample to enter the counter, which is surrounded by a glass tube to form an annular space for the liquid radioactive sample.

In the author's experience the life of end window halogen quenched counters is limited by damage to the end window, which is quite fragile, rather than by internal failures.

Examination of the construction of commonly used counters such as the Mullard MX148 shows the rather surprising feature of an anode that is several millimetres in diameter. This at first sight appears to contradict what has already been said about the need for very high electric fields for gas multiplication of the primary electrons. Closer examination shows that the surface of the tungsten anode is quite rough. When it is considered that the electric field close to a spherically tipped point is proportional to the inverse of the radius squared it becomes obvious that locally high electric fields can exist at many points on the surface of the anode and so multiplication is possible.

A further feature of halogen quenched counters is that they operate at considerably lower voltages than the older organic quenched counters, typically in the range 400–500 V instead of 1000–1500 V. This is again an advantage for portable equipment as it simplifies further the high voltage supply. In addition halogen quenched tubes seem to be more tolerant of accidental over voltages.

5

....

The scintillation counter

5.1   Introduction

The scintillation counter is comprised of two main components: firstly, a scintillator that absorbs incident radiation and converts the energy deposited by ionisation into a fast pulse of light and, secondly, a photomultiplier. This second component converts the light pulse into a pulse of electrons and also amplifies the electron pulse by a very large factor by means of a sequence of secondary emission stages. Further external amplification is generally necessary before the pulse can be processed or recorded.

5.2   Scintillators

The main requirements for a scintillator are high efficiency for detection of the desired radiation, a high light output, a short decay time for the emission of light so that fast counting is possible, and negligible after-glow. Many different materials are used for scintillators, depending upon the application, and these can be classified into the following main groups: alkali–halide crystals, organic crystals, organic liquid solutions, plastic solid solutions and certain glasses. Table 5.1 lists the main parameters for some common commercially available scintillators together with their main applications.

5.2.1   Alkali–halide scintillators

Sodium iodide is the most widely used scintillator for detection of gamma-radiation and the way in which alkali–halide crystals produce light will be illustrated by reference to sodium iodide. Normally, large single crystals are used for scintillators because of their uniformity and superior light transmission properties, but polycrystalline sodium iodide with good light transmission properties is also available. A crystal of pure sodium iodide is not suitable for use as a scintillator. When an ionising particle loses energy in the scintillator it excites electrons from the valence band across the forbidden gap to the conduction band. These electrons rapidly return to the vacancies left in the valence band and in so doing they can cause emission of quanta of light in the ultra-violet region of the spectrum, as shown in Fig. 5.1(*a*). Light of this wavelength, around 200 nm, would be strongly absorbed by the glass window in front of the photocathode of the photomultiplier and, in addition, would be strongly absorbed by the sodium iodide itself since the quantum energy matches the band gap energy. The emitted light can therefore be absorbed by raising an electron across the gap and, since radiationless transitions are also likely to occur, returning the electron to the valence band without emission of radiation, the number of quanta escaping from the scintillator

Table 5.1. *Some commonly used scintillators*

Scintillator	Type	Use	Peak λ (mm)	Decay constant (ns)	Relative light output
NaI(Tl)	crystal	X, γ	413	230	1.000
CsI(Tl)	crystal	γ, heavy particles	580	1100	0.413
LiI(Eu)	crystal	n	475	1200	0.326
ZnS	crystalline	α	450	200	1.304
Anthracene	organic crystal	α, β, γ, fast neutrons	447	30	0.435
Stilbene	organic crystal	γ, fast neutrons	410	4.5	0.217
NE213	organic liquid	fast neutrons	425	3.7	0.339
NE102A	plastic	α, β, γ, fast neutrons	423	2.4	0.283
NE908	glass	slow neutrons	399	75	0.070

Reproduced by permission of Nuclear Enterprises Ltd, Edinburgh.

will only be small. It is therefore necessary to create quanta of visible light in order that they may reach the photocathode with minimal losses. To achieve the production of visible light a few percent of thallium is included in the solution from which the crystal is grown so that thallium impurities occur within the crystal. The purpose of these impurities is to create intermediate energy levels within the forbidden gap. Electrons returning from the conduction band can now break their return at these intermediate energy levels and so emit quanta of light of lower energy and longer wavelength, as shown in Fig. 5.1(b):

$$\lambda = 1224/\Delta E(\text{eV}) \text{ nm} \tag{5.1}$$

Thallium is chosen as the activator for sodium iodide since it produces energy levels with a spacing of ΔE in the region of 3 eV, therefore resulting in photons of about 3 eV energy of 400 nm wavelength. This process is not very efficient as only one photon of energy about 3 eV is produced for roughly every 50 eV lost by the ionising radiation. This means that for 1 MeV absorbed in the scintillator only about 20 000 photons at the lower end of the visible spectrum are produced.

Fig. 5.1. Electron energy levels in (a) pure and (b) thallium doped sodium iodide.

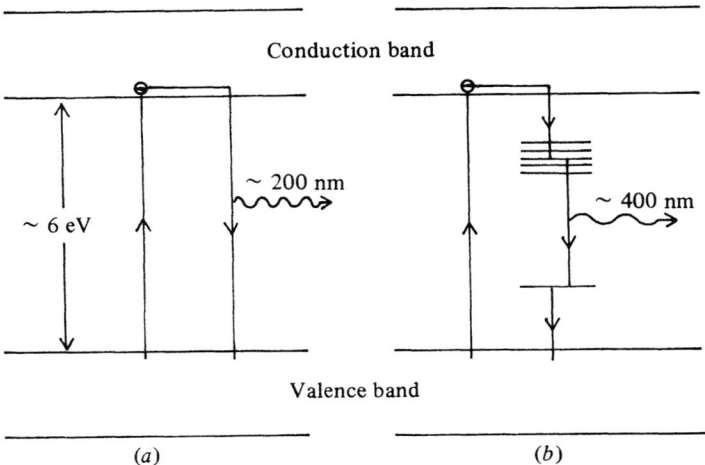

Sodium iodide has the disadvantage that it is hygroscopic and deliquescent and so it must be protected from atmospheric moisture. The scintillation crystals are therefore sealed in a metal can, usually of aluminium which is a weak absorber of gamma-radiation, and the can has a flat glass window at one end to enable the light to escape from the scintillator to the photomultiplier. A diffuse reflector, commonly made of magnesium oxide, fills the space between the can and the scintillator crystal, apart of course for the crystal face in contact with the window where an optical grease is used to improve the light transmission. Crystal sizes depend on the application and on the energy of the X or gamma-radiation to be detected. For low energy X-rays in the region of 10 keV the crystal is only a few millimetres thick since the absorption coefficient is very high for sodium iodide at this energy range and the front face is covered with a thin beryllium window ($Z = 4$) to minimise X-ray absorption before reaching the scintillator. For gamma-ray detection crystal sizes range from a (length × diameter) of 15 mm × 15 mm for lower energy applications up to 300 mm × 300 mm for high energy gamma-radiation.

Higher atomic number atoms have a larger photoelectric cross-section than lower atomic number atoms, as described in Chapter 1. In sodium iodide the iodine ($Z = 53$) is mainly responsible for the photoelectric absorption that results in effectively complete absorption of the gamma-ray energy and hence to a light pulse whose amplitude is a measure of the gamma-ray energy. However, Compton scattering and pair production can also be important absorption processes and these give rise to light pulses of lower amplitude than the photoelectric interaction. In Fig. 5.2 the mass absorption coefficient of sodium iodide is shown as a function of gamma-ray energy and the photoelectric and pair production components are also shown, enabling the contributions of the three interaction mechanisms to be seen.

The light output from the scintillator falls with time from the beginning of the pulse in an exponential manner that is characterised by a decay constant τ, the time for the light to decay to $1/e$ of its previous value:

$$I = I_0 \exp(-t/\tau) \qquad (5.2)$$

In some types of scintillator it is necessary to describe the light decay by several decay constants such that the light intensity is expressed as a sum of several exponentials with different amplitudes. For sodium iodide the principal decay constant is 230 ns, which means that 98% of the light is emitted within 1 μs of

Fig. 5.2. Mass absorption coefficient of sodium iodide as a function of gamma-ray energy. (Calculated from data in *Nuclear Data Tables A7*, 565–681, Photon cross-sections from 1 keV to 100 MeV for elements $Z = 1$ to $Z = 100$, E. Storm, H. I. Israel, Academic Press (1970).)

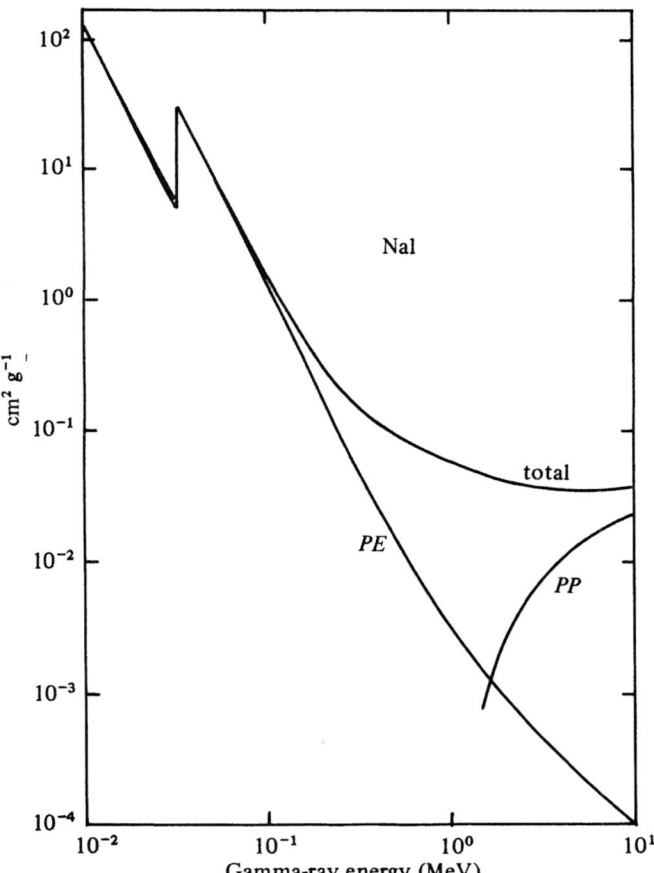

the ionising interaction. A high counting rate is therefore possible, but it may need to be limited to minimise pile-up of the random pulses if distortion of energy information is to be avoided. This effect may be aggravated when the pulse is shaped by external amplifiers.

Ionisation is caused by fast electrons produced by gamma-ray interactions in sodium iodide. These electrons have a low linear energy transfer or low ionisation density along their path and so the light emission process is not saturated and, except at vey low gamma-ray energies, the light intensity per pulse may be regarded as proportional to the energy deposited in the scintillator. This is important for gamma-ray energy spectroscopy which will be described in Section 5.4.

5.2.2   Organic scintillators

With the exception of ZnS(Ag) and glass, all the remaining scintillators listed in Table 5.1 are organic in either solid or liquid form. They all contain aromatic hydrocarbon molecules that have suitable vibrational excited states which can be excited by ionising radiations and which will de-excite by emission of electromagnetic radiation. Anthracene and stilbene are pure organic crystals and are among the earliest scintillators. Liquid scintillators, many of which are commercial compounds, are principally used for internal counting of radioactive solutions or for fast neutron detection. They are not pure substances but typically comprise about 95% of a solvent such as toluene, xylene or dioxan, about 5% of a primary activator such as PPO (2,5 diphenyl-oxazole) and about 0.5% of a wavelength shifter such as POPOP (1,4-di(2-(5-phenyl-oxazolyl))). The exact mixtures of the commercial compounds are not necessarily made available by the manufacturers. In these scintillators the ionisation process transfers energy to the solvent since this is the predominant component and the energised solvent then excites the activator which emits quanta of ultra-violet radiation. This ultra-violet radiation is then absorbed by the wavelength shifter which reradiates it at a longer wavelength at the lower end of the visible spectrum.

The liquid scintillator NE213, which is based on the solvent xylene and also contains naphthalene as well as an activator and POPOP wavelength shifter, has the interesting property, like stilbene, of having different light decay constants when excited by fast neutrons (recoil protons) and gamma-rays (electrons). By means of suitable electronics, as discussed in Section 5.5, it is possible to distinguish between these two types of radiation, a valuable feature since neutrons are generally accompanied by gamma-radiation.

Plastic scintillators such as NE102A are very useful for alpha-particle and beta-particle detection. Since they contain only the light atoms hydrogen and carbon the photoelectric cross-section is very low compared with the Compton cross-section and for gamma-radiation below about 2 MeV only Compton interactions are observed. Figure 5.3 shows the mass absorption coefficient for NE102A as a function of gamma-ray energy, together with the contributions of photoelectric and pair production interactions. Compare this with Figure 5.2. Another feature of organic scintillators, particularly the plastic types is that they have a very short decay constant which enables them to be used for timing the flight path of fast neutrons and so find their energy. Although plastic scintillators are sensitive to gamma-radiation they do not possess pulse shape discrimination. If neutrons are produced with the aid of a pulsed nuclear accelerator it may be possible to reject gamma-ray events because they will have a different time of arrival at the detector compared with fast neutrons. These scintillators are a solid solution of a primary activator and a wavelength shifter in an aromatic plastic such as polyvinyltoluene.

5.2.3 Glass scintillators

Glass scintillators can be made from lithium silicate glass activated with cerium. Since they contain a substantial proportion of lithium they are useful as slow neutron detectors, ionisation being produced by the charged alpha-particle and triton or hydrogen-3 nucleus released in the neutron reaction:

$$^6Li + n \rightarrow {}^3H + {}^4He + 4.8 \text{ MeV}$$

This reaction has a high cross-section for slow neutrons, being 946 barns at the reference energy of 0.0253 eV, equivalent to a neutron velocity of 2200 ms⁻¹. Natural lithium only contains 7.56% of ^{6}Li, the remainder being ^{7}Li that does not undergo a charged particle reaction with slow neutrons. Glasses containing lithium enriched or depleted in ^{6}Li are available and the scintillator NE908, listed in Table 5.1, contains about 7% of ^{6}Li and very little ^{7}Li.

Fig. 5.3. Mass absorption coefficient of NE102A as a function of gamma-ray energy. (Calculated from data in *Nuclear Data Tables A7*, 565–681, Photon cross-sections from 1 keV to 100 MeV for elements $Z = 1$ to $Z = 100$, E. Storm, H. I. Israel, Academic Press (1970).)

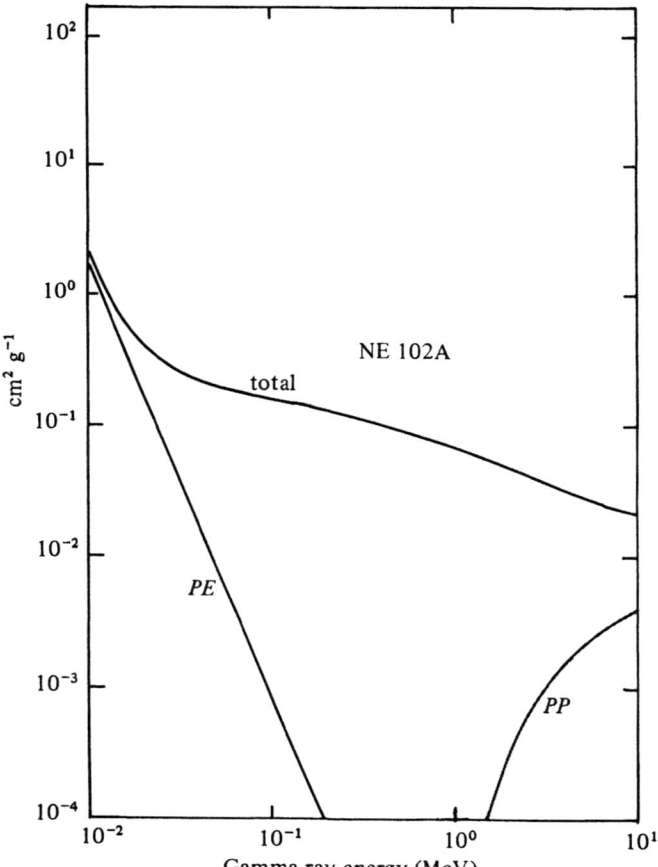

5.2.4 Other scintillators

Zinc sulphide, activated with silver, is a possible detector for alpha-particles when deposited as a thin layer. It mainly finds application in a dual phosphor for measurement of alpha-particle and beta-particle contamination in which a thin layer of ZnS(Ag) is deposited on a sheet of plastic scintillator with the ZnS(Ag) facing the source. Large pulses are produced by alpha-particles in the zinc sulphide and smaller pulses by beta-particles in the plastic.

5.3 The photomultiplier

There are several designs of photomultiplier but all have the following common features:

(a) A photoemitting cathode that converts the light from the scintillator into a shower of electrons. Typical photo-sensitive materials are Cs_3Sb-O, $Na_2KSb-Cs$ and $RbCs$ and these form a thin layer on the inside of the photo-multiplier window.

(b) A series of electrodes at progressively greater positive potentials with respect to the cathode. These electrodes, generally known as dynodes, have secondary emitting surfaces that release several electrons for each electron accelerated from the previous stage that strikes it. A common dynode material is beryllium copper. Electron multiplications of around ×4 per dynode are typical but this is dependant on the acceleration voltage from the previous stage. Overall multiplications of up to 10^8 are possible in 14 stage tubes and the widely used 11 stage tubes typically have a multiplication of 10^6.

(c) An anode to collect the multiplied electrons.

(d) An evacuated glass envelope to contain the electrode structure.

(e) An external resistance chain to act as a potential divider for applying the appropriate potentials to each of the electrodes. The actual values of the resistances in this chain depend on the application for which the complete scintilla-tor is to be used and must be specified by the designer or user.

In Fig. 5.4 an 11 stage photomultiplier having a venetian blind style of dynodes is shown schematically together with a possible resistance chain. Connections are indicated opposite the relevant electrodes for clarity but in reality they emerge from the tube on a multi-pin base at the opposite end of the glass envelope to the cathode. The capacitors shown across the later stages of multiplication help to maintain a constant voltage across these stages during the electron pulse, which can draw a high current over a very short time interval, and this helps to maintain linearity of amplification for differing pulse sizes. Values of the unit of resistance R shown in Fig. 5.4 may vary between 0.1 MΩ and 3 MΩ according to application. For low counting rates a high value may be used and this results in lower electrical noise but for high counting rates the value of R must be low such that the signal current is only a small fraction of the steady current in the resistance chain. Sometimes a high voltage zener diode is used in place of the resistance between cathode and the first dynode. This then maintains a constant voltage over the first stage when it is necessary to adjust the overall

Fig. 5.4. Photomultiplier tubes: Venetian blind, box and grid and focussed types.

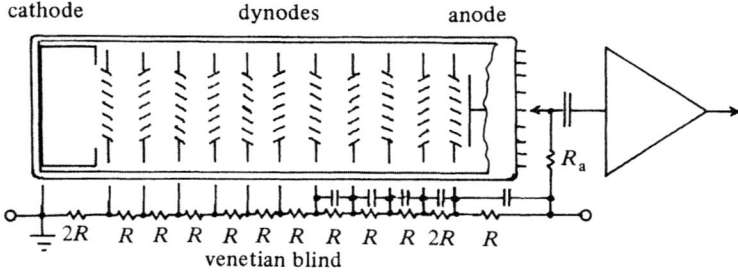

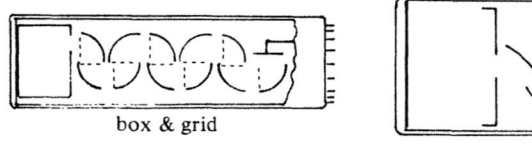

voltage across the device to alter the gain of the system and so helps to maintain efficient collection of electrons from the cathode by the first dynode.

Also shown in Fig. 5.4 are two other types of dynode structure: the box and grid type that is used in smaller diameter tubes (venetian blind tubes are usually used for cathode diameters of 50 mm or more), and the focussed dynode structure that is used in tubes for fast timing applications. This latter type is designed to have a very low spread in electron arrival time at the anode, typically of only 2–3 ns. Cathode diameters in commonly available photomultiplier tubes range from 19–130 mm but cathodes up to 300 mm diameter can be made. Up to 130 mm diameter the cathode window is optically flat and so is easily coupled to the scintillator, but larger diameter cathodes need a convex window to withstand the differential pressure and so are mainly suitable for use with liquid scintillators. In tubes having a cathode diameter greater than 50 mm the dynodes and anode are contained in a neck of about 50 mm diameter.

The electron multiplication obtained is a function of both the voltage between stages and the total number of stages or dynodes, and hence is also a function of the overall voltage between cathode and anode. The gain over an individual stage is not quite linear with voltage. Over a limited range of overall voltage V the total gain can be expressed approximately by the expression

$$\text{gain } G \propto V^n$$

where n is of the order of 9 for an 11 stage tube. This index n is not quite constant with applied voltage and also shows considerable batch variations for a particular type of tube. Since the gain varies as a high power of the applied voltage it is essential to use a highly stabilised power supply if consistent results are to be obtained. The fractional change in gain is related to the fractional change in voltage by

$$\delta G/G = n\delta V/V$$

so for a typical 11 stage tube a 1% voltage change would result in a 9% change in gain. For spectroscopic (energy meaurement) applications a voltage stablity of better than 1 part in 10^4 is needed and this requires both a stable voltage supply and high stability resistors in the voltage divider chain. With a typical voltage per stage of around 75 volts and a cathode to first dynode potential of at least 150 volts, power supplies must be able to give of the order of 1000 volts and a current of a few milliamps when a low resistance chain is used.

Despite the high internal amplification obtained in a photo-multiplier tube further external amplification of the signal is generally necessary. The value of the anode load resistance R_a is determined by the distance of the amplifier from the detector. Normally a pre-amplifier is mounted close to the tube and an anode load R_a of 0.1–1.0 MΩ is suitable as it only has to act as a return to the high voltage. If, however, a long coaxial cable has to be used to connect the detector to the amplifier then the anode load should be made equal to the characteristic impedance of the cable, which is most usually 50 Ω. The load resistor then acts as a termination for the cable and so prevents multiple signal reflections that could lead to distortion of the pulses. Electron transit times from cathode to anode are of the order of 50 ns but the spread in time of arrival of electrons at the anode depends upon the electron optical properties of the dynode structure. For a sharp electron pulse leaving the cathode the time spread at the anode is about 20 ns for venetian blind dynodes and 2–3 ns for focused dynodes. This latter type is therefore suitable for use with fast decay time scintillators when used to measure the time of arrival of fast ionising particles.

5.4   Gamma-ray spectroscopy

There are many applications for scintillation counters but the most widely used one is gamma-ray spectroscopy, which is the measurement of the energy and intensity of gamma-radiation. For this application the scintillator is sodium iodide NaI(Tl) and for gamma-rays emitted from the common radioactive emitters cylindrical crystals with a length × diameter in the range 38 mm × 38 mm–76 mm × 76 mm are useful.

As described in Chapter 1 the processes by which gamma-rays can transfer energy to the scintillator are:

(1) The photoelectric effect, in which virtually all of the gamma-ray energy is transferred to the crystal.

(2) Compton scattering, in which there is partial transfer of energy to the crystal, ranging from zero to a maximum of

$$E_c = E_\gamma/(1 + m_0c^2/(2E_\gamma))$$

(3) Pair production, which has a threshold energy of 1.022 MeV giving a total energy of $(E_\gamma - 1.022)$ MeV to the electron and positron. When the positron comes to rest and is annihilated by any free electron, two gamma-rays of 0.511 MeV each are released and one or both of these gamma-rays *may interact in the crystal*. This process is negligible around the threshold but becomes significant at about 2 MeV for sodium iodide.

It will be observed that only the photoelectric process deposits the whole of the gamma-ray energy in the crystal. Since the gamma-ray interactions produce fast electrons and these have a low *linear energy transfer or low ionisation density per unit length* of path, the amount of light released is proportional to the electron energy. As a consequence the photoelectric interaction produces a quantity of light (number of photons) proportional to the gamma-ray energy. For NaI(Tl) about 20 000 photons of wavelength around 400 nm are released per MeV absorbed in the crystal. The spread in the number of these photons is larger than would be expected from simple statistical considerations. It is affected by the optical homogeneity of the crystal, the variation in effectiveness of the diffuse reflector surrounding the sides and back of the crystal and the region in the crystal where the gamma-ray interacts. For gamma-radiation around 1 MeV in energy the standard deviation in the number of photons is about 2.5% for a good crystal and can be considerably larger for a poor crystal. The standard deviation does not vary very much with gamma-ray energy. Note that if the standard deviation is taken to be the square root of the number of photons then 20 000 photons would have a standard deviation of only 0.7%.

Since the conversion efficiency of photons to electrons at the photocathode is typically of the order of 10% (although some tubes are considerably better than this) it is only to be expected that the electron population will also exhibit significant fluctuations. The electron population will typically have a standard deviation similar in magnitude to the photon population but this will depend on gamma-ray energy, as will be shown later in this section.

So far only the photoelectric effect has been considered. To see the effect of the other interactions consider the experimentally observed pulse height distributions at different gamma-ray energies. The typical spectrum produced by a gamma-ray of 1 MeV is shown in Fig. 5.5. Since at this energy the only possible reactions for gamma-ray absorption are the photoelectric effect and Compton scattering the pulse height spectrum has a photopeak or full energy peak corresponding to the complete absorption of a gamma-ray and also a lower pulse height continuous distribution resulting from the electrons produced in Compton scatters. In addition there is usually a low pulse height subsidiary peak that

Fig. 5.5. Typical pulse height spectrum produced by 1 MeV gamma-rays incident on a sodium iodide detector.

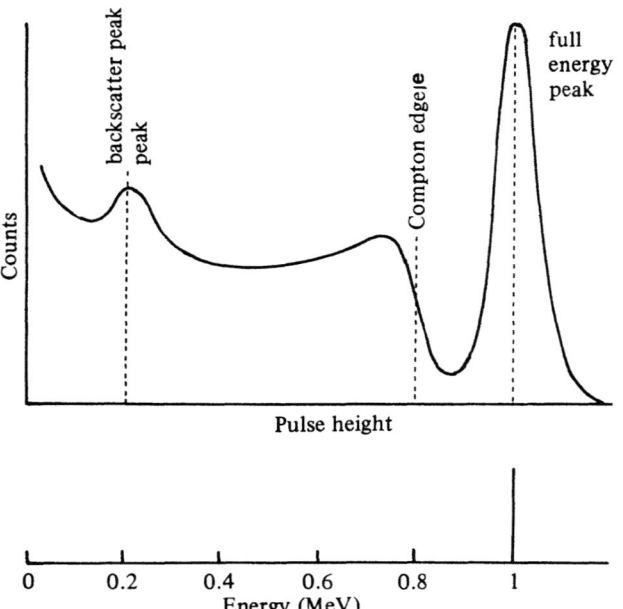

results from gamma-rays that have been Compton backscattered at around 180° from the source holder or surrounding shielding and this corresponds to an energy of $(E_\gamma - E_c)$. The broadening effect of fluctuations in the number of photons and of photoelectrons on the pulse height distribution can be seen in Fig. 5.5. The term 'full energy peak' is more correct than photopeak since it is possible to build up a full energy event by a sequence of reactions. For example, the scattered gamma-ray from a Compton interaction may then be absorbed in a photoelectric interaction, the two events being so close together in time that the photons appear to the photomultiplier to have been produced at the same time. Multiple processes are more likely in larger crystals and in these the full energy peak is larger, compared with the Compton distribution, than in small crystals.

The typical spectrum produced by a gamma-ray of 2 MeV is shown in Fig. 5.6. At 2 MeV pair production is becoming a significant reaction (see Fig. 5.2). The pulse height spectrum still

Fig. 5.6. Typical pulse height spectrum produced by 2 MeV gamma-rays incident on a sodium iodide detector.

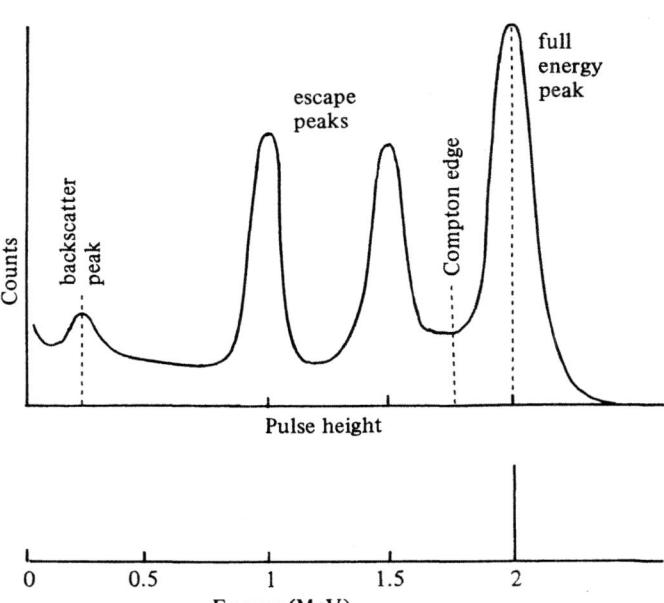

shows a full energy peak, a backscatter peak and a Compton distribution but there are now two further peaks, known as 'escape peaks', corresponding to energies of $(E_\gamma - 1.022)$ MeV where both annihilation gamma-rays have escaped from the crystal and $(E_\gamma - 0.511)$ MeV where one of the annihilation gamma-rays has been completely absorbed and the other has escaped. At this energy range a further contribution to the full energy peak is obtained if both annihilation gamma-rays are completely absorbed.

The pulse height of the photopeak is found to be a linear function of gamma-ray energy down to about 20 keV. Below this energy the ionisation density along the electron path increases rapidly with decreasing electron energy, and linearity fails due to saturation of the light emitting centres. For all practical gamma-ray energies the scintillator counter may be regarded as having a linear calibration of photopeak pulse height against gamma-ray energy and may be conveniently calibrated using standard long lived radioactive emitters such as [137]Cs (0.66 MeV) and [60]Co (1.17 and 1.33 MeV).

Because what should ideally be a very sharp full energy peak is broadened by the effects of photon and electron fluctuations, there is a limit to the detail that can be seen in complex spectra. An

Fig. 5.7. Summation of two Gaussian peaks with separations of 0.5, 1.0, 1.5, 2.0 FWHM.

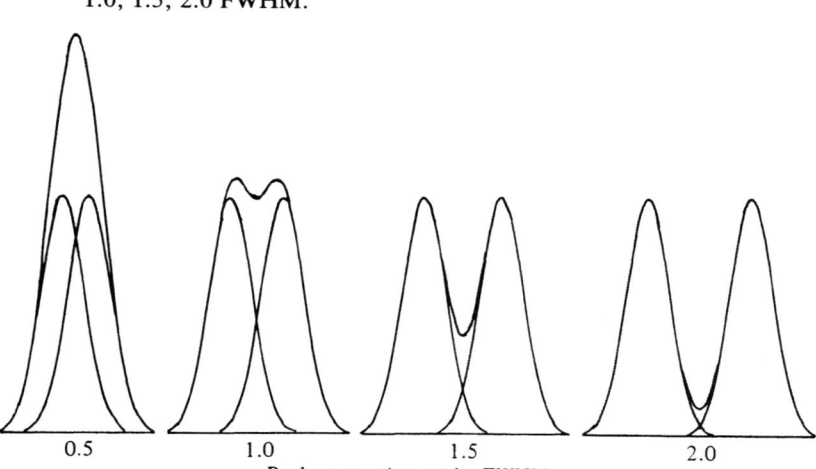

0.5   1.0   1.5   2.0
Peak separation–units FWHM

isolated full energy peak is closely Gaussian in shape and a useful quantity to characterise the peak is its full width measured at half the vertical height of the peak, the *Full Width at Half Maximum height or FWHM*. Fig. 5.7 shows the effect of summing two identical Gaussian peaks that are separated by 0.5, 1.0, 1.5 and 2.0 FWHM. It will be seen that for a separation of 1.0 FWHM the valley between the peaks is almost filled and at 0.5 FWHM the two peaks add together to look like one peak. Conventionally two peaks are said to be resolvable if their separation is greater than the FWHM. A useful qualitative check of a system is provided by examining the spectrum from ^{60}Co whose two gamma-rays differ in energy by 12.8% of the mean value. If the resolution of the system is good then there will be a deep valley between the two peaks.

Although the photon standard deviation σ_p for the scintillator can only be measured rather than defined, it is possible to quantify the electron standard deviation in the photomultiplier tube. Assume a simple model in which N electrons leave the cathode, the first stage has a gain G_1 and all the remaining stages have a gain G. Table 5.2 shows the number of electrons and their standard deviation at various stages of the photomultiplier. The overall standard deviation for the electrons is found by taking the square root of the sum of the squares of the individual standard deviations.

$$\sigma_e^2 = 1/N + 1/NG_1 + 1/NG_1G + 1/NG_1G^2 + 1/NG_1G^3$$
$$+ \cdots + 1/NG_1G^{10}$$

for an 11 stage photomultiplier.

Table 5.2. *Electron multiplication in a photomultiplier*

Stage	Number of electrons	Standard deviation (fractional)
cathode	N	$1/\sqrt{N}$
dynode 1	NG_1	$1/\sqrt{(NG_1)}$
dynode 2	NG_1G	$1/\sqrt{(NG_1G)}$
dynode 3	NG_1G^2	$1/\sqrt{(NG_1G^2)}$
dynode 4	NG_1G^3	$1/\sqrt{(NG_1G^3)}$
dynode 11	NG_1G^{10}	$1/\sqrt{(NG_1G^{10})}$

Therefore

$$\sigma_e^2 = \frac{1}{N} \left[1 + \frac{1}{G_1} \left(\frac{1 - 1/G^{11}}{1 - 1/G} \right) \right] \tag{5.3}$$

Since typically G is about 4 then $1/G^{11}$ may be neglected in comparison with unity and the electron standard deviation reduces to

$$\sigma_e = \sqrt{\frac{1}{N} \left(1 + \frac{G}{G_1(G - 1)} \right)} \tag{5.4}$$

The overall standard deviation including both the electron and photon contributions then becomes

$$\sigma = \sqrt{(\sigma_e^2 + \sigma_p^2)}$$

and for a Gaussian shaped peak the FWHM is 2.36 standard deviations. Therefore the FWHM or resolution, expressed as a function of the mean full energy peak size, is

$$\text{FWHM} = 2.36\sqrt{(\sigma_e^2 + \sigma_p^2)} \tag{5.5}$$

Since the number of photoelectrons N is proportional to the gamma-ray energy it is possible to write the electron standard deviation for fixed gain conditions as

$$\sigma_e^2 = a/E_\gamma$$

Also σ_p^2 is approximately constant with energy and will be written as a constant b to obtain

$$\sigma^2 = a/E_\gamma + b$$

or

$$(\text{FWHM})^2 = 2.36^2(a/E_\gamma + b) \tag{5.6}$$

Consider by way of example a gamma-ray of 1 MeV yielding $N = 2000$ electrons with $G_1 = G = 4$ and $\sigma_p = 2.5\%$. The electron standard deviation is therefore 2.6% and the total standard deviation is 3.6%, giving FWHM = 8.3% at 1 MeV.

Fig. 5.8 shows how the resolution would vary with gamma-ray energy for a detector having the characteristics assumed in the above example.

5.5   Pulse shape discrimination

Certain organic scintillators, such as stilbene and the commercial liquid scintillator NE213, have different light decay constants when excited by fast neutrons or by gamma rays, as already mentioned in Section 5.1.2. The principal decay constant of the emitted light is affected by the ionisation density along the path of the charged particle that excites the emission of light, and the slow light decay is stimulated by a high ionisation density. Hence recoil protons that are generated in the scintillator by fast neutron interactions will be associated with the slow component of light decay and electrons, which are near the minimum ionisation density in the MeV range, will be associated with the fast component of light decay. Several methods have been developed for distinguishing between the two decay rates and hence for particle identification, thus enabling both gamma-radiation and fast neutrons to be detected and identified using one detector. Two such methods will be described, the zero crossing method and the charge comparison method.

5.5.1 Zero crossing method

The zero crossing method of pulse shape discrimination can be effected using standard commercially produced electronics. It requires two signals from the photomultiplier, one a linear signal

Fig. 5.8. Variation of resolution with gamma-ray energy for a typical sodium iodide detector.

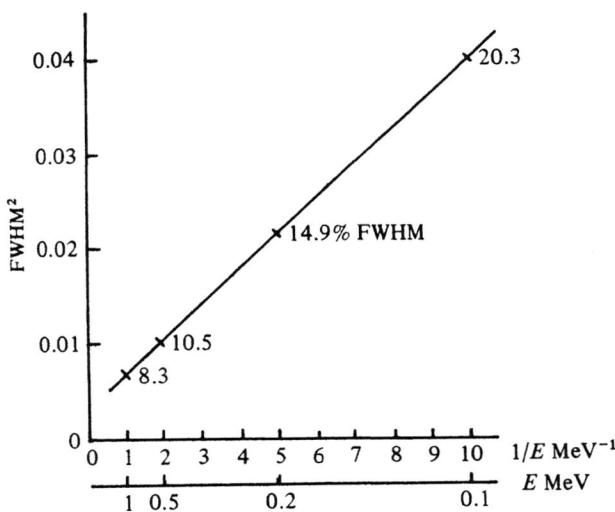

(i.e. proportional to light output) that conveys both decay rate information in its shape and energy information in its amplitude and a timing signal for use in measuring the pulse shape of the linear signal. It is usual to derive the linear signal from one of the later dynodes, usually the 10th dynode in a 14 stage photomultiplier tube, which needs to be of the focussed type to minimise the effect on pulse shape of the spread in electron transit time. Unlike an anode signal the dynode signal will be positive since more electrons leave a dynode for the next stage than arrive at it. This dynode pulse is amplified by an integrating pre-amplifier and then by a main amplifier that produces a bipolar output pulse (see Chapter 7 for a description of amplifiers and pulse shaping). The anode signal from a 14 stage photomultiplier tube will be considerably larger than the linear pulse from the 10th dynode and will generally not be proportional to the light output from the scintillator. If the photomultiplier tube is operated with the anode supply at earth potential and the cathode at negative potential then the anode pulse can be developed across a resistor whose resistance is equal to the characteristic impedance of the output cable and the negative anode pulse can be fed directly to a fast discriminator that provides a fast negative logic pulse from the leading edge of the anode signal. Fig. 5.9 shows the photomultiplier connections needed for these signals. The light decay time somewhat affects the shape of the bipolar pulse from the amplifier and the time from start to when the voltage crosses zero is longer for the slow light decay. It is only a small effect, typically 100 ns difference between fast and slow decay zero crossing times for an average time from start to zero crossing of 2500 ns (2.5 μs) but this is sufficient for discrimination. A zero crossing discriminator that gives a standard fast negative logic pulse when the input pulse goes through zero provides information about the decay time or particle type by comparing the coincidence of this pulse with the suitably delayed timing signal from the anode. A coincidence resolving time of 100–150 ns is needed to sort out either fast neutron or gamma-ray induced signals, the particular radiation being selected by the amount of delay chosen for the anode pulse. Fig. 5.9 also shows the time relation of the various signals and indicates a typical time spread in the zero crossing signal obtained in a mixed radiation field with a range of fast neutron and gamma-ray energies.

5.5.2 Charge comparison method

The zero crossing method needs very careful setting up and is not effective over a wide dynamic pulse range unless great care is

Fig. 5.9. Zero crossing method of pulse shape discrimination: photomultiplier connections; bipolar pulse shapes; signal timings; time spread of zero crossing signal.

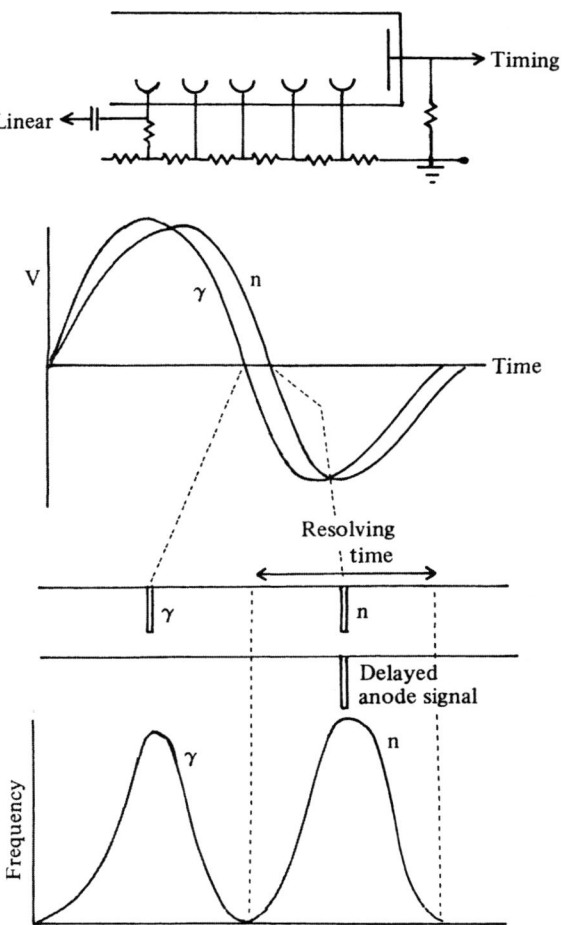

taken to eliminate any DC levels on the linear pulse. Even then the time from start to zero crossing tends to be dependent on pulse size and can shorten for large proton pulses. The comparison method developed at AERE Harwell and available as a commercial unit (LINK 5010), uses only the anode pulse with the photomultiplier running at a low enough voltage such that the output pulse is linearly proportional to light output from the scintillator. The anode is directly coupled to two integrating circuits operating in parallel. One of these integrates for 25 ns and then holds the result, and the other integrates for 500 ns, these timings being suitable for the liquid scintillator NE213. By comparison of the two integrator outputs a decision can be made electronically as to whether the pulse was induced by a neutron or a gamma-ray (see Fig. 5.10). The system provides an identifying pulse that can be counted or used to open the input gate of a multi-channel analyser so that the amplified linear signal can be recorded. Provided that the photomultiplier used is a low noise type the system is capable of handling a wide

Fig. 5.10. Charge comparison method of pulse shape discrimination. Integrator outputs as a function of time for gamma-ray and neutron signals.

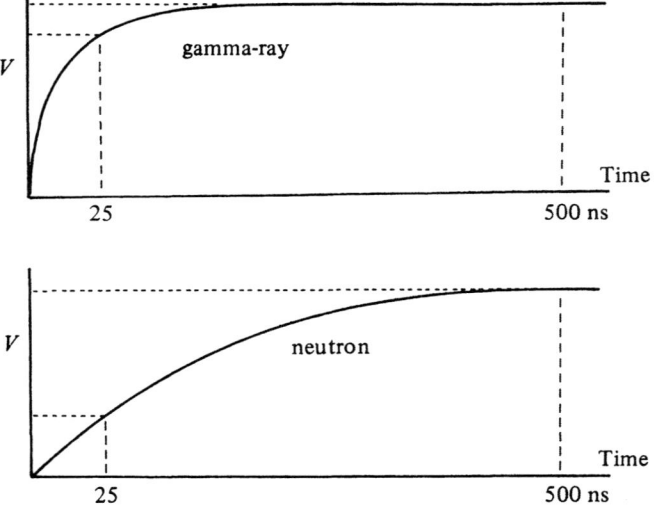

dynamic range of pulses and it is claimed that under suitable conditions pulses from neutrons down to about 50 keV can be recognised.

Certain other scintillators apart from NE213 exhibit pulse shape discrimination. CsI(Tl) can distinguish between alpha particles and gamma-radiation. A dual scintillator comprising a sandwich of two scintillators of different decay times can also be used. The first layer would be thin to detect short range radiations such as protons or alpha-particles, and the second layer would be thick to detect penetrating radiation such as gamma-rays. One such suitable combination could be CsI(Tl) followed by NaI(Tl).

5.6   Burst cartridge detection

In the Magnox natural metallic uranium fuelled and gas cooled power reactors there is continuous monitoring of the carbon dioxide gas coolant stream to see if any fission products are present which would indicate a burst fuel can. It is essential to discover such a burst quickly since metallic uranium rapidly oxidises in hot CO_2 and severe contamination of the reactor could result if the damaged fuel were left in the reactor. There are between 3000 and 6000 fuel channels in a Magnox reactor, depending on the design, and the gas flowing through these is normally sampled in groups until a can failure is detected, the groups being sampled in sequence with an overall scan time of about one hour. When a suspect group is found the individual channels in the group can then be scanned to identify the channel containing the faulty fuel element. Depending on the severity of the release of radioactivity a decision can then be taken to discharge the fuel channel at once or wait until the next convenient maintenance period.

A sample of CO_2 gas is taken from a channel or group of channels and passed to a counting system that is arranged to monitor only certain isotopes of krypton and xenon of which the two most important are the fission products ^{90}Kr and ^{140}Xe. These two isotopes decay as follows:

$$^{90}_{36}Kr \xrightarrow[33s]{\beta^-} {}^{90}_{37}Rb \xrightarrow[2.7m]{\beta^-} {}^{90}_{38}Sr$$

$$^{140}_{54}Xe \xrightarrow[10s]{\beta^-} {}^{140}_{55}Cs \xrightarrow[66s]{\beta^-} {}^{140}_{56}Ba$$

The Kr and Xe fission products, being gaseous, will be carried through to the counting system with the CO_2 and will decay to the solids Rb and Cs. These are removed from the gas stream by electrostatic precipitation onto a charged wire at $-4\,kV$, precipitation being continued for a period of 30 s for an individual sample. The wire is then indexed into a beta sensitive scintillator counter that will detect any significant amounts of radioactive Rb or Cs. Counting continues for 30 s before the wire is indexed out of the counter and replaced by the next sample length. A long continuous loop of wire is used so that about half an hour elapses before the section of wire is reused, thus allowing any activity collected to decay to insignificant proportions. Fig. 5.11 shows the main features of a burst cartridge detection system.

5.7   Cerenkov detectors

Charged particles that travel faster than the phase velocity of light in a transparent medium will cause the emission of quanta of light. This phenomenon is known as the 'Cerenkov effect' and is a process distinct from the scintillation effect. Since the phase velocity c is c_0/n where c_0 is the velocity of light in free space,

Fig. 5.11. Burst cartridge detection system for a gas-cooled nuclear reactor.

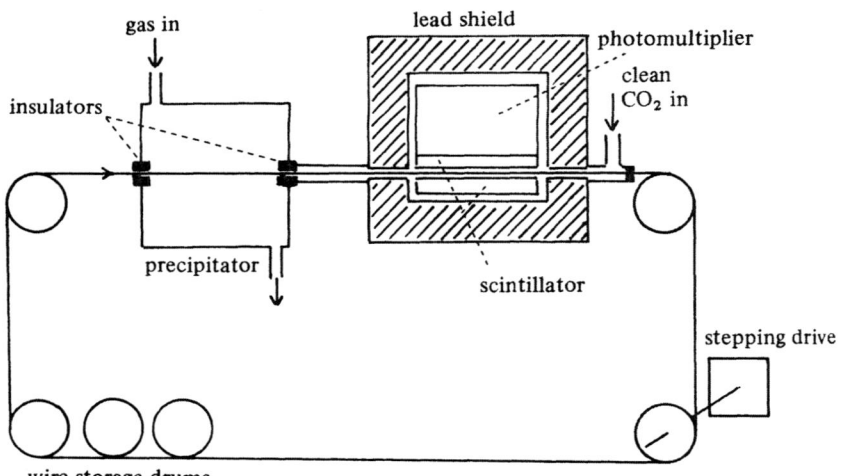

3×10^8 m s^{-1}, and n is the refractive index of the medium, then the effect is restricted to moderate to high energy electrons (the velocity of a 1 MeV electron is $0.866c_0$), or to protons of energies around 1 GeV or greater as produced by very high energy accelerators, or to very energetic cosmic ray particles. From relativistic considerations the minimum energy E_0 of an electron that can produce light in a medium of refractive index n is given by

$$E_0 = m_0c^2(n/\sqrt{(n^2 - 1)} - 1) \qquad (5.7)$$

and if gamma-radiation produces the electrons, the minimum gamma-ray energy is found by equating the energy of the Compton electron associated with a gamma-ray scatter of 180° to the minimum electron energy E_0. This yields a minimum gamma-ray energy for production of light of

$$E_{\gamma_0} = \frac{m_0c^2}{2}\left(\frac{n + 1}{\sqrt{(n^2 - 1)}} - 1\right) \qquad (5.8)$$

The threshold energies for production of Cerenkov radiation by electrons (E_0) and gamma-rays (E_{γ_0}) as a function of refractive index are shown in Fig. 5.12.

Fig. 5.12. Threshold energy for production of Cerenkov radiation as a function of refractive index for electrons and gamma-rays.

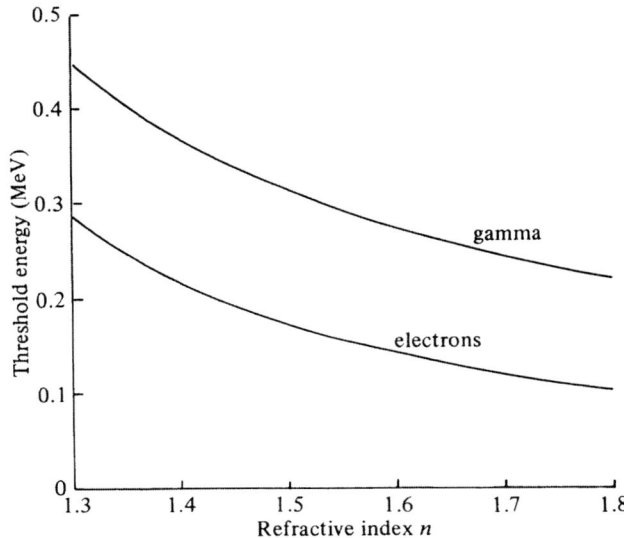

The advantages of Cerenkov counters are that it posseses a threshold of energy detection and so discriminates against low energy radiation, and the pulse of light emitted is very much faster than a scintillation pulse since it is emitted during the time taken for the electron to slow down to threshold energy. Its main disadvantage is the low light output which is only about $\frac{1}{100}$th of the light output from a high efficiency scintillator.

The emitted light is directional and is confined to a forward pointing cone of semiangle $\cos^{-1} (c_0/vn)$ about the particle direction, where v is the particle velocity. The spectral distribution of the emitted light is proportional to $1/\lambda^2$, so an ultra-violet sensitive photomultiplier will increase the size of the recorded signal.

5.8   Gas scintillation counters

Some high purity gases, notably xenon and helium, have been used as scintillators. Due to the low stopping power of gases they are only suitable for detection of short range intensely ionising particles such as alpha-particles or fission fragments. Light emission is simply a direct result of ionisation and excitation produced by passage of the charged particle, and the light is mainly in the ultra-violet. Consequently, either a photomultiplier that is sensitive to ultra-violet, generally with a thin coating of an ultra-violet to visible wavelength converter on the window, must be used or a wavelength shifter, such as a trace of nitrogen, must be added to the gas. Scintillation efficiency is low, at best less than 10% of the light emission from sodium iodide for the same amount of energy deposited in the scintillator. Enhanced efficiency has been obtained by including electrodes in the gas scintillator such that it operates like a proportional counter. The aim of this modification is not to collect the charge released by multiplication but to enhance the light output by increasing the number of ionising events associated with each pulse. Provided that the counter is operated within the proportional region, linearity of light output with energy deposited is maintained.

Further reading

J. B. Birks, *The Theory and Practice of Scintillation Counting*, Pergamon, 1964.

G. W. McBeth, J. E. Lutkin & R. A. Winyard, 'A simple zero crossing pulse shape discriminator system', *Nucl. Instr. & Meth.*, **93**, 99–102, 1971.

J. M. Adams & G. White, 'A versatile pulse shape discriminator for charged particle separation and its application to fast neutron time-of-flight spectrometry', *Nucl. Instr. & Meth.*, **156**, 459–76, 1978.

V. V. Verbinski, W. R. Burrus, T. A. Lowe, W. Zobl, N. W. Hill & R. Textor, 'Calibration of an organic scintillator for neutron spectrometry', *Nucl. Instr. & Meth.*, **65**, 8–25, 1968.

D. Slaughter & R. Strout, 'Flyspec: a simple method of unfolding neutron energy spectra measured with NE213 & stilbene spectrometers', *Nucl. Instr. & Meth.*, **198**, 349–55, 1981.

6

·· ··· ··· ··· ··· ··· ··· ··· ··· ··· ··· ··· ··· ··· ··· ··· ··· ··· ··· ···

Semiconductor detectors

6.1   Introduction

When ionising radiations pass through any material, gaseous, liquid or solid, ionisations are created as energy is lost. Solid materials have the advantage of high density and hence high stopping power, especially compared with gases, and so offer the prospect of a compact and high efficiency detector. If a high electric field can be sustained across the solid then the charge released by the ionisation can be collected and the presence of ionising radiation so recorded. It is essential that the current flow through the detector in the absence of ionising radiations is negligible in order that the small quantity of charge that is produced by the ionising event can be observed, and so metals are totally unsuitable. The ionisation process in crystalline solids consists of raising electrons from the valence band to the conduction band, a process that leaves an equal number of positive holes in the valence band. Materials with a small band gap such as germanium (0.7 eV) and silicon (1.1 eV) require an average of 3.0 eV and 3.6 eV, respectively, to be expended in creating a hole–electron pair, and so the amount of primary charge released is about ten times that in a gas-filled detector. One serious difficulty with germanium is that thermal excitation of electrons across the band gap gives pure germanium a high conductivity at room temperature. As a consequence any detector using germanium has to be cooled to liquid nitrogen temperature to reduce the current flow to sufficiently low values in order to observe the ionisation current pulse. Cooling of silicon detectors is generally only necessary when they are used for detection of low energy ionising radiations such as characteristic X-rays.

Various constructional methods are used for semiconductor detectors, depending upon their application, but they are based on the principle of the reverse biased rectifying junction. The two main semiconductor materials, silicon and germanium, are in group four of the periodic table. Pure materials are known as intrinsic semiconductors and have equal numbers of holes and electrons, where the carrier density is dependent on the width of the band gap E_g and the absolute temperature T:

$$n_i = p_i \propto \exp\left(-E_g/2kT\right) \tag{6.1}$$

where n is the electron carrier density, p is the positive hole carrier density and the subscript i indicates that the material is intrinsic.

Semiconductors of lower resistivity can be manufactured by the addition of selected doping impurities during crystal growth. Addition of a group five material, such as phosphorus, creates impurity levels known as donor levels just below the bottom of the conduction band (see Fig. 6.1). The extra valence electron from an impurity atom or donor atom is readily released by thermal excitation into the conduction band. At room temperature virtually

Fig. 6.1. Electron energy levels in intrinsic, n-type and p-type semiconductors.

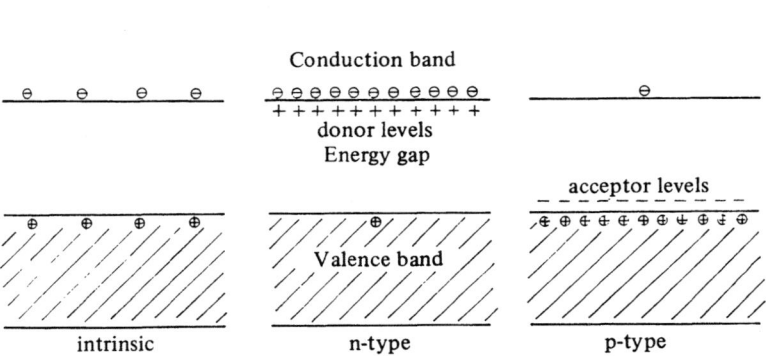

all the donor atoms are ionised and the conductivity is proportional to the donor atom concentration. The donor concentration N_d is very small. Even a concentration of one part per million produces a high conductivity (low resistivity) material. Since thermal excitation across the gap is a dynamic process with continual excitation and recombination setting up an equilibrium, the hole density is much reduced because the high electron density in the conduction band compared with intrinsic material greatly increases the recombination rate with the positive holes left in the valence band by thermal excitation of electrons. Hence electrons are the majority carriers and, consequently, this type of semiconductor is known as n-type (for negative majority carriers). At normal temperatures the carrier densities can be expressed as

$$n_n = N_d \quad \text{and} \quad n_n p_n = n_i^2 \tag{6.2}$$

where $n_n \gg n_i$.

If the added impurity is a group three material such as boron then one bond is left unsatisfied at each impurity atom position. The incomplete bond produces energy levels termed 'acceptor levels' just above the top of the valence band (see Fig. 6.1) and electrons from the valence band can easily be thermally excited into these acceptor levels where they remain trapped. As a result of removing these electrons from the valence band an equal number of positive holes is created and at room temperature effectively all of the acceptor levels are occupied and the hole density is equal to the acceptor density $p_p = N_a$. The probability of electrons in the conduction band recombining is greatly increased by the high hole density and so the density of electrons in the conduction band is much reduced compared with intrinsic material. Consequently holes are the majority carrier and the material is designated p-type (for positive holes majority carrier). In a similar manner to n-type material,

$$n_p p_p = n_i^2 \tag{6.3}$$

where

$$n_p \ll n_i$$

6.2   Silicon surface barrier detector

The earliest type of semiconductor detector that is still widely used is the Silicon Surface Barrier Detector or SSB. Essentially this is a reverse biased p–n rectifying junction and before describing the construction and operation of such a device the characteristics of a reverse biased junction will be discussed.

A p–n junction is created when the doping impurity type suddenly changes with distance through the crystal to give a transiton from p-type to n-type material. If the p-side is connected to the negative of a voltage source and the n-side is connected to the positive then the junction is reverse biased and only a very low current will flow in the reverse direction to current in a forward biased junction. Reverse biasing draws the majority carriers away from the junction region and so leaves a net negative charge density on the p-side of the junction due to the ionised acceptor atoms and a net positive charge density on the n-side of the junction due to the ionised donor atoms. These fixed charge densities oppose the removal of the mobile hole and electron charge carriers by an external voltage and the removal of the majority carriers only extends over a limited region, called the depletion layer, whose width depends on the applied voltage and on the concentration of donor and acceptor atoms. The width of the depletion layer can be found by solving Poisson's equation for a region of fixed charge density. In one dimension Poisson's equation can be written as

$$d^2V/dx^2 = -\rho/\varepsilon \qquad (6.4)$$

Solution of this equation shows that the depletion layer width is proportional to the square root of the voltage across the junction region and inversely proportional to the square root of the doping atom density,

$$W \propto \sqrt{(V_j/N_d)} \qquad (6.5)$$

The voltage across the junction is not quite equal to the external voltage since a small voltage, of the order of one volt in silicon, is set up by diffusion of charge carriers even when no external voltage is applied. Provided that the external bias voltage is several volts it is usually sufficient to regard the reverse voltage across the junction as being equal to the external applied voltage. Since the regions of semiconductor outside the depletion layer have a fairly high conductivity and the reverse current flowing through the device is generally very much less than a microamp, the external voltage is effectively applied to the depletion region. The voltage distribution across the depletion layer follows a square law with respect to distance from each edge of the depletion layer. It is possible to create depletion layers with widths in the range 100–1000 μm and the electric field strengths in the depletion layer can be up to 10^6 Vm^{-1} without breakdown, even with quite modest applied voltages. Hence if ionisation is produced within the depletion layer, that is, the production of equal numbers of extra electrons and holes by the action of an ionising particle, then the holes will be swept to the p-side and the electrons to the n-side by the action of the internal field, and the charge pulse can be detected by suitable amplification, just as for a gas-filled ionisation chamber.

Depletion layers greater in depth than 1000 μm are difficult to obtain. By reference to Fig. 1.6, which gives the range energy relationship for various charged particles in silicon, it will be seen that such a device used as a detector will only be able to detect short range particles such as alpha-particles. In addition, the response to gamma- or X-radiation will be negligible due to the shallow sensitive region and the low attenuation coefficient of silicon with its low atomic number.

For any charged particle it is essential to have a very thin entrance region such that negligible energy is lost before the particle enters the depletion region. The silicon surface barrier detector is designed to accomplish this and its construction is shown in Fig. 6.2. A wafer of n-type silicon is lightly oxidised on the polished upper surface and then a thin layer of gold is evaporated onto the oxide layer in order to form an ohmic contact. The electrical properties of the oxide layer are such that it forms a rectifying junction with the n-type material and the gold plus oxide layer is sufficiently thin so that charged heavy particles can cross it with very little energy loss. On reverse biasing the depletion layer that is formed is therefore almost completely in the n-type region. Only low bias voltages are needed to create depletion layers sufficiently thick to stop alpha-particles, and detectors will work, although with incomplete charge collection even with zero bias although this is not recommended. It is easy to exceed the reverse breakdown voltage and the bias voltage should always be applied and removed smoothly, for example with the aid of a suitable R–C circuit to avoid transients that could exceed the breakdown voltage. Silicon surface barrier detectors may be operated in vacuum but to avoid damaging discharges caused by the high field across the depletion layer the bias voltage should not be applied during pumping down.

Energy resolution obtained from such devices is not as good as might be expected from theoretical considerations. Take, as an example, an alpha-particle of 5 MeV energy. The average number of charge carriers of either sign that is produced by complete stopping of the alpha-particle is

$$5 \times 10^6/3.6 = 1.39 \times 10^6$$

Assuming that the Fano factor is 0.1 then the standard deviation in the number electrons or holes is

$$\sqrt{(1.39 \times 10^6 \times 0.1)}, \text{ which is equal to } 373$$

Hence the ideal full width at half maximum would be

$$2.36 \times (373/(1.39 \times 10^6)) \times 5000 \text{ keV} = 3.17 \text{ keV}$$

Fig. 6.2. Silicon surface barrier (SSB) detector.

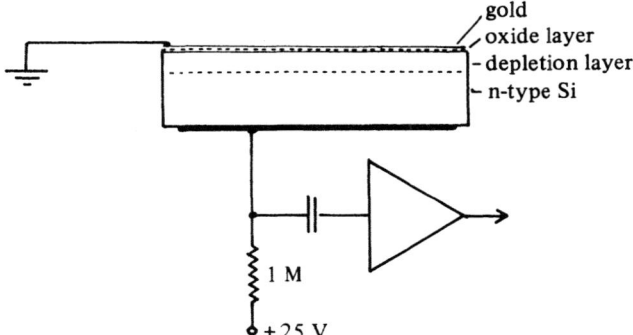

gold
oxide layer
depletion layer
n-type Si

1 M

+25 V

The actual performance of these devices is much worse than this, 25 keV FWHM being about the best obtainable for an alpha-particle. Electrical noise is responsible for the broadening and is comprised of three components:

(i) fluctuations in the bulk reverse leakage current, an effect that can be reduced by cooling;
(ii) fluctuations in surface leakage current, an effect that depends on fabrication techniques and usage of the detector;
(iii) Johnson noise which is due to the series resistance of the undepleted region and poor electrical contacts to the detector.

Overall the FWHM is a function of ionisation effects and noise $(FWHM)^2 = (FWHMionisation)^2 + (FWHMnoise)^2$.

6.3   Lithium drifted silicon detectors

Detectors with sensitive depths of several millimeters are needed for the detection of higher energy beta-particles and of X-rays. As has already been seen there is a limit to the depth of the depletion layer in a silicon surface barrier detector since too high a bias voltage will cause reverse breakdown. An alternative approach is to create a junction with an intrinsic (undoped) layer between the p- and n-layers. The intrinsic region then forms the sensitive region and the p- and n-layers are effectively the contacts to the sensitive region. A bias voltage has to be applied to attract the electrons released in ionisation to the n-layer and the holes to the p-layer and is much larger than the bias for the silicon surface barrier detector since the sensitive region is now much thicker.

Fabrication of a p–i–n junction with a thin entrance window for charged particles is carried out as follows. A single crystal of p-type silicon a few millimetres thick has its front face coated with lithium in an inert atmosphere to prevent oxidation of the lithium. Under the action of heat and an electric field the lithium atoms can be made to diffuse or drift through the crystal structure to the desired depth of the intrinsic layer. Since lithium is an alkali metal and has one valence electron it can easily give up the valence electron to neutralise an acceptor site and so prevent holes being created in the valence band. The effect is to turn the layer over which the lithium has been drifted into electrically intrinsic region by neutralisation of the acceptor atoms. The lithium is not drifted all the way through the crystal so that a p-type region remains. Connection is made to this via an evaporated aluminium contact. At the front surface there will be excess lithium, thus producing a thin region in which the majority carrier will be electrons and this can be regarded as a thin n-type region. Connection is made to this n-region with a thin evaporated gold layer. Fig. 6.3 shows the general arrangement of *such a device which is commonly known as a Si(Li) detector*. A polarising voltage such that the p-region is negative with respect to the front face is applied and this is typically in the range 500–2000 V, depending on the thickness of the sensitive layer (the intrinsic region).

At room temperature the Si(Li) detector may be used for the detection of beta-particles. Due to their energy spread a high energy resolution is not important but for detection of characteristic X-rays the device must be cooled to reduce the component of FWHM that is due to electrical noise. As has already been seen, in

Fig. 6.3. Lithium drifted silicon (Si(Li)) detector.

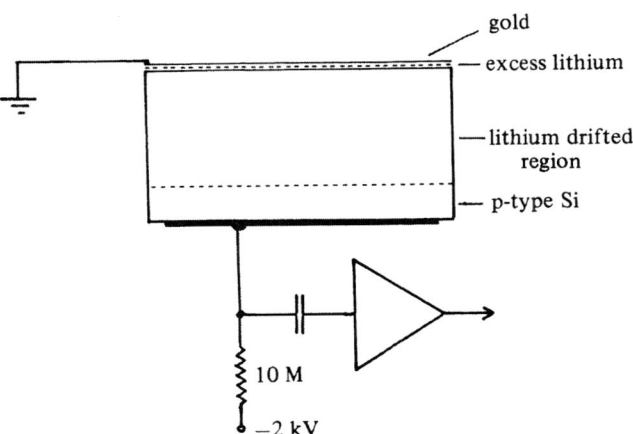

intrinsic material the number of charge carriers varies exponentially as the inverse of the absolute temperature (equation (6.1)),
so cooling silicon from about 300 K (room temperature) to about
90 K (liquid nitrogen temperature) causes a drastic reduction in the
magnitude of the reverse leakage current, which forms a large
component of the noise. When a detector is cooled it has to be in a
vacuum chamber to insulate it and prevent frosting. Consequently
a thin entrance window made of beryllium ($Z = 4$) has to be
included as part of the wall of the vacuum chamber to permit
X-rays to enter with only low losses by absorption. In Fig. 6.4 the
efficiency of detection of X-rays as a function of X-ray energy is
shown for a typical Si(Li) detector, including the absorbing effect
of the beryllium window. Since for Cu(K_α) X-rays (energy
8.04 keV) the average number of charge carriers of either sign
produced is only $8040/3.6 = 2233$, a very low noise pre-amplifier is
also needed and it is usual to cool the input transitors of the
pre-amplifier (generally FET's) by combining the detector and

Fig. 6.4. Efficiency of detection as a function of X-ray energy for a
5 mm thick Si(Li) detector with a 0.025 mm thick beryllium entrance
window. (Reproduced by permission of Canberra Instruments Ltd,
Swindon.)

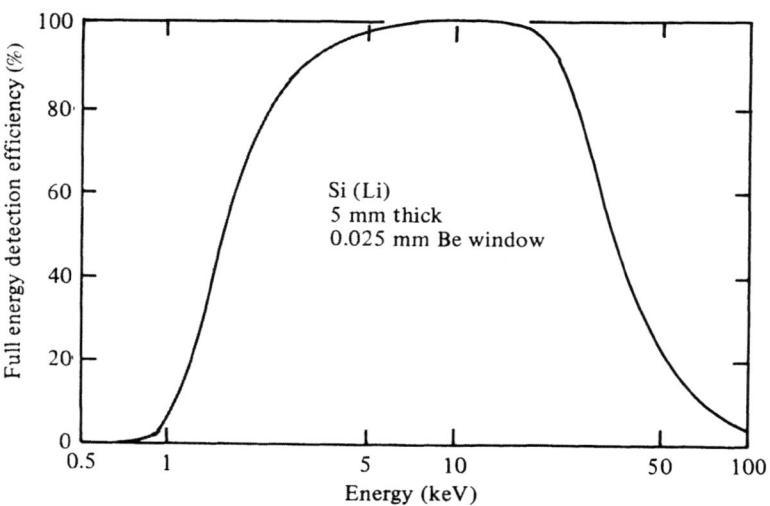

Si (Li)
5 mm thick
0.025 mm Be window

preamplifier into a single unit that fits onto the dewar of liquid nitrogen.

With a Fano factor of about 0.11 the theoretical FWHM for $Cu(K_\alpha)$ radiation is

$$2.36 \times \sqrt{(8040 \times 3.6 \times 0.11)} = 133 \text{ eV}$$

for a perfectly noise free system. It proves impossible to eliminate all noise but a good detector can have a resolution at this energy of about 170 eV, indicating a noise contribution of 106 eV (this is mainly amplifier noise). The K_α X-ray is really a doublet and the separation of the $K_{\alpha 1}$ and $K_{\alpha 2}$ components increases with increasing atomic number and hence energy. Fig. 6.5 shows the separation of the $Tb(K_\alpha)$ lines (44.482 keV and 43.744 keV) as well as the $Cu(K_\alpha)$ line for which the doublet (8.048 keV and 8.028 keV) cannot be resolved.

6.4   Germanium detectors

As can be seen from Fig. 6.4 the efficiency of detection of electromagnetic radiation for Si(Li) detectors is very low above 100 keV. Since photoelectric absorption is roughly proportional to Z^4/E_γ^3 to Z^5/E_γ^3, depending on E_γ germanium with $Z = 32$ will be a much more efficient detector at higher energies than silicon which

Fig. 6.5. Pulse height spectrum for copper and terbium K X-rays obtained with a Si(Li) detector.

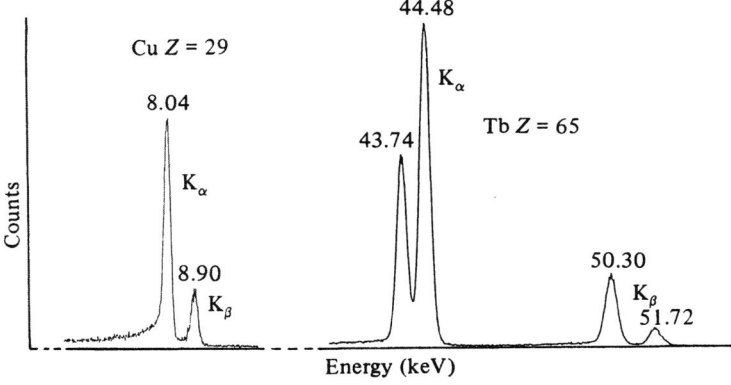

has $Z = 14$. Consequently, any gamma-ray detector must be fabricated from germanium, which even in its intrinsic form has a low resistivity at room temperature. Any germanium detector, therefore, must be operated at liquid nitrogen temperature in order to reduce the electrical noise sufficiently to obtain detectable signals of good resolution. Gamma-ray detectors do not necessarily need a thin entrance window and a widely used form of the detector is the coaxial lithium drifted germanium detector or Ge(Li). This is fabricated from a cylindrical p-type crystal of germanium of large volume, typically 50 cm^3 or greater. Lithium is coated onto the cylindrical surface and drifted towards the centre under the action of heat and an electric field to create an effectively intrinsic region surrounding a cylindrical p-type core. As the surface region contains excess lithium this will act as an n-type region. Fig. 6.6 shows the general layout of a cylindrical Ge(Li) detector. It is essential to maintain cooling at all times on such detectors or otherwise the lithium will diffuse out of its location. A good detector will have a resolution of around 3.5 keV at a gamma-ray energy of 1 MeV, which is considerably greater than the figure of 1.3 keV which is calculated using 3.0 eV per ionisation and a Fano factor of 0.1, so electrical noise is still a significant factor. Nevertheless the resolu-

Fig. 6.6. Lithium drifted germanium (Ge(Li)) detector.

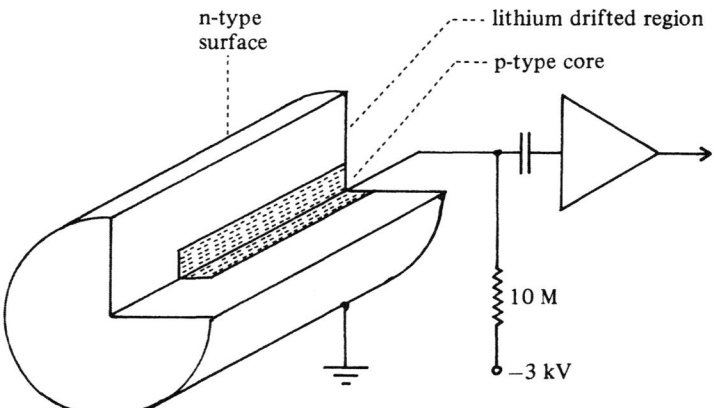

tion obtained is about 20 times better than for a NaI(Tl) scintillation counter. Fig. 6.7 shows the pulse height distribution in the
region of the two full energy peaks for ^{60}Co obtained with a Ge(Li)
detector.

More recent developments have lead to a detector fabricated
from high purity (intrinsic) germanium. The planar versions are
generally suitable for detecting the higher energy characteristic
X-rays from high atomic number elements, as well as gamma-rays,
since they have a thin entrance window which will allow the X-rays
to enter. The contact at the entrance window face of the detector is
a thin layer of p-type material which is most reliably manufactured
by ion implantation of boron into the surface layer. An n-type layer
can be produced on the reverse face by a limited region of lithium
drifting. High purity germanium can never be completely impurity
free and is generally slightly p-type. On application of a voltage to
reverse bias the device the weak p–n junction depletion layer
increases in width with applied voltage until it extends to the thin

Fig. 6.7. Typical pulse height spectrum for ^{60}Co in the region of the
two photopeaks for a Ge(Li) detector.

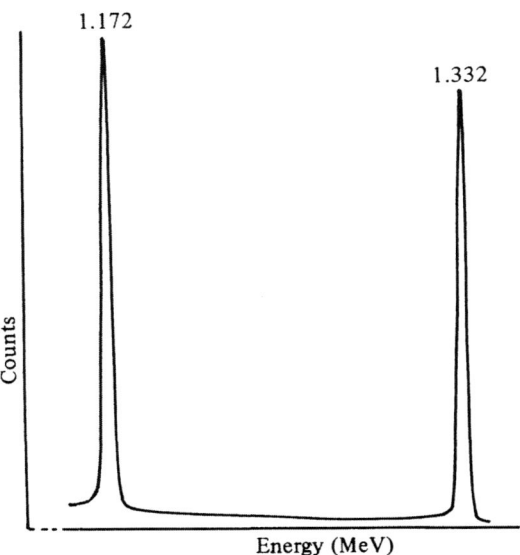

heavily doped p-type surface layer. This is possible with the very high resistivity material produced by cooling without breakdown since the width of the depletion layer varies as the square root of the resistivity as well as the square root of the applied voltage. Total depletion is therefore possible and the sensitive depth is almost equal to the full thickness of the device. Since the sensitive region does not contain lithium the detector can be stored at room temperature when not in use. Planar detectors are commercially available attached to small liquid nitrogen dewars that contain enough liquid nitrogen for about 12 hours use and these can be hand held as they are much more compact than a Ge(Li) detector with its large nitrogen dewar. Resolution is generally similar to that obtained with Ge(Li) detectors.

7

... .

Electronics for nuclear detectors

7.1   Introduction

This chapter does not attempt to describe detailed circuits but is intended to deal with the main principles behind the most common electronic modules for processing the output from radiation detectors. There are many carefully developed modules that are commercially available and these are sufficient to satisfy the needs of most experimentalists. In the most usual form these consist of NIM (*N*uclear *I*nstrumentation *M*odule) units that plug into a bin containing all the necessary power supplies. A standard NIM bin will accommodate 12 single width modules.

7.2   Amplifiers

With the exception of the Geiger–Mueller counter all detectors produce very small pulses that must be amplified before further processing, such as counting or pulse size analysis, can be carried out. An amplifier will be regarded as a 'black box' that is a high gain system with externally applied feedback in order to obtain a controlled and reproducible gain. The basic symbol used in circuit diagrams for such an amplifier is shown in Fig. 7.1 and the

Fig. 7.1. Symbolic representation of an amplifier.

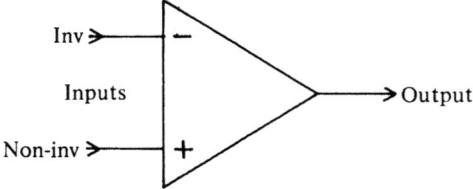

open loop gain of the system (without feedback) is generally of the order of 10^4, or greater, which can often be regarded as being infinite for circuit analysis. Most common operational amplifiers are unsuitable for use in nuclear instrumentation since the gain–frequency characteristic starts to fall away rapidly at frequencies well below 1 MHz. A few operational amplifiers can work at frequencies greater than 1 MHz and are sometimes found in equipment intended mainly for use with sodium iodide scintillation detectors, which have a comparatively long light decay constant (see Table 5.1). For amplifiers that have the highest linearity of pulse size between input and output, and which can handle the pulses from fast scintillators, discrete components are used (separate transitors, diodes, resistors and capacitors) and this is generally also necessary for DC coupled systems. Whatever the construction the high open loop gain amplifier can still be regarded as a 'black box' for the purpose of understanding the basic principles of feedback amplifiers.

Feedback is necessary in order to obtain stable amplification as the open loop gain of apparently identical systems can vary significantly due to tolerances in components. Two classes of amplifiers used in nuclear instrumentation can be identified: the voltage sensitive amplifier and the charge sensitive amplifier; and these differ in the type of feedback that is used. Voltage sensitive amplifiers use resistive feedback and the principle of an inverting amplifier is shown in Fig. 7.2. Since the amplifier has a very high

Fig. 7.2. Amplifier with resistive feedback, inverting output.

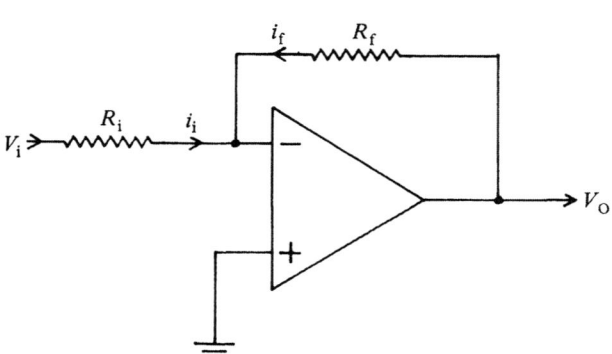

open loop gain the two inputs only require a very small voltage difference between them for a large output voltage. For example, if the open loop gain is 10^4 then a 1 mV difference between the inputs will produce a 10 V output. Hence in the circuit of Fig. 7.2 both inputs will always be very close to earth (a virtual earth). This implies that the currents in the input and feedback lines must be almost identical in magnitude but of the opposite sign since the input impedance of the amplifier will be high between the two inputs and the input current will be very low due to the small potential difference between the two inputs. If an infinite open loop gain is assumed then the input current

$$i_i = V_i/R_i$$

will be equal and opposite in sign to the feedback current

$$i_f = V_o/R_f$$

i.e.

$$V_o/R_f = -V_i/R_i$$

and therefore

$$V_o = -V_i(R_f/R_i) \tag{7.1}$$

The effective gain is therefore R_f/R_i, which may only be a factor of 10. High gains are obtained by cascading several stages rather than using one high gain stage since the frequency response is improved and the system is less affected by component variations. Gain is not the only requirement of an amplifier for pulse processing. The shaping of the pulse for matching of units is also important and this will be discussed later in this chapter.

It will be noticed that the amplifier that has just been described using resistive feedback inverts the output with respect to the input. An alternative arrangement for resistive feedback that is non-inverting is possible and is shown in Fig. 7.3. Again the potential difference between the two inputs can be regarded as negligible due to the high open loop gain so that the fraction of the output voltage that appears at the inverting input is effectively equal to the input voltage that is applied to the non-inverting input:

$$V_o R_2/(R_1 + R_2) = V_1$$

∴

$$V_o = V_i((R_1 + R_2)/R_2) \tag{7.2}$$

with this configuration the gain of the system has to be greater than unity.

The other amplifying system that is in common use is the integrating amplifier. This is more usually used as a pre-amplifier but can also be obtained as a complete scintillation counter amplifier that needs no pre-amplifier, and is generally designed for use with a sodium iodide scintillator. The basic arrangement of an integrating amplifier is shown in Fig. 7.4. If it is assumed initially that the feedback resistance R_f is infinite in value then it can be ignored and any current in the feedback loop is directly related to the rate of change of output voltage:

$$q = C_f V_o$$

∴

$$i = \frac{dq}{dt} = C_f \frac{dV_o}{dt}$$

Just as for the first example of resistive feedback the net current at the input can be set to zero such that

$$C_f \frac{dV_o}{dt} + i_i = 0 \tag{7.3}$$

where $i_i = V_i/R_i$

∴

$$V_o = \frac{1}{C_f} \int i_i \ dt = \frac{q_i}{C_f} \tag{7.4}$$

Fig. 7.3. Amplifier with resistive feedback, non-inverting output.

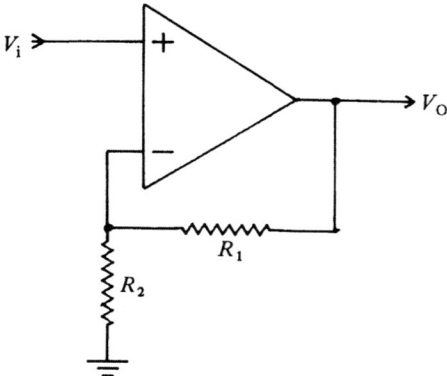

Since most detectors produce a small quantity of charge that is related and often proportional to the energy deposited by the ionising radiation in the detector, then the output voltage from an integrating amplifier is also a measure of the energy deposited. If, however, this system were used for recording a succession of pulses the output voltage would rise in steps until the amplifier output saturated. To overcome this a finite resistance in parallel with the feedback capacitor has to be used to allow the charge to leak away between pulses, and the actual response obtained will then depend on the shape and duration of the input pulses as well as on the component values in the feedback loop. When the feedback resistance is included the differential equation to be solved contains one more term than equation (7.3) to allow for current through the feedback resistor.

$$C_f \frac{dV_o}{dt} + \frac{V_o}{R_f} + i_i(t) = 0$$

or

$$\frac{dV_o}{dt} + \frac{V_o}{C_f R_f} + \frac{i_i(t)}{C_f} = 0 \qquad (7.5)$$

Let $C_f R_f = \tau_f$ which is the time constant of the feedback loop. Two cases are of interest. The first is simply to regard the input pulse as being a short rectangular pulse of total charge $q_i = i_i T$ where the time duration of the pulse, T, is very much less than the time

Fig. 7.4. Integrating amplifier (capacitive feedback).

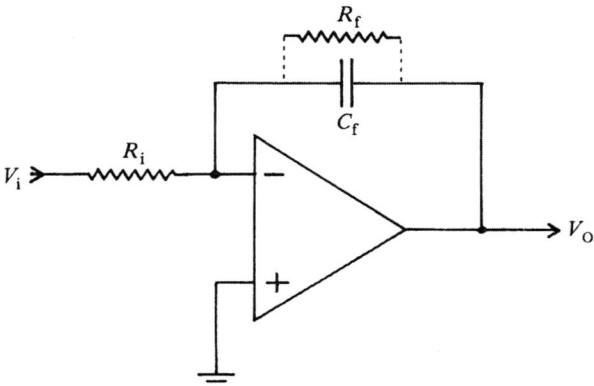

constant τ_f. In this case the capacitor charges to a voltage of q_i/C_f and then decays exponentially with a time constant τ_f. The output voltage thus decays exponentially with time:

$$V_o = \frac{q_i}{C_f} e^{-t/\tau_f} \tag{7.6}$$

This expression is an approximation for fast detectors such as plastic scintillators but will not hold for slower detectors such as proportional counters or NaI(Tl) scintillation counters where the light decays in a time comparable with τ_f. For the NaI(Tl) scintillation counter the light output and hence the anode current in the photomultiplier decays exponentially with a time constant of about one-quarter of a microsecond. In this case the input current can be written as

$$i_i(t) = i_o \exp(-t/\tau)$$

and equation (7.5) becomes:

$$\frac{dV_o}{dt} + \frac{V_o}{C_f R_f} + \frac{i_o}{C_f} e^{-t/\tau} = 0 \tag{7.5a}$$

This can easily be solved either directly or by use of the Laplace transform to yield

$$V_o = \frac{-q_i \tau_f}{C_f(\tau_f - \tau)} (e^{-t/\tau_f} - e^{-t/\tau}) \tag{7.7}$$

where

$$q_i = \int i_i(t)\, dt = i_0 \tau$$

Note that if $\tau = \tau_f$ then equation (7.7) has to be expressed as

$$V_o = \frac{-q_i t}{C_f \tau_f} e^{-t/\tau_f} \tag{7.7a}$$

If $\tau_f \gg \tau$ then this approximates to the case of a short rectangular input pulse. In Fig. 7.5 the shape of the output pulse is shown for $\tau_f = \tau$ and $\tau_f = 25\tau$.

The main advantage of the integrating amplifier is that the effective input capacity is $C_f \times$ (open loop gain) and this value is generally much higher than the capacity of cabling used in the input circuit plus the stray capacitance of the detector to earth. As a result the size of the output signal is not greatly affected by the length of input cable. On the other hand voltage feedback amplifiers show a marked drop in gain as the input cable length is increased, particularly as detectors commonly have high output impedances.

7.2.1 Pre-amplifiers

The usual way of overcoming the problem of signal attenuation when it is necessary to use long cables between the detector and the amplifier is to use a pre-amplifier close to the detector and a long length of cable between the pre-amplifier and the main amplifier. The pre-amplifier has a dual purpose. Firstly, it amplifies the small detector pulse (by about 10^2 for semiconductor and gas-filled detectors but only by unity for scintillation detectors) and, secondly, it provides a low impedance output that is capable of driving the signal into a long length of cable with only a small loss of amplitude.

Two types of pre-amplifier are in common use. One is either a non-shaping or short time constant amplifier that produces an amplified pulse that closely follows the shape of the input signal. The other is an integrating type with τ_f typically 25 μs, a time

Fig. 7.5. Output pulse from an integrating amplifier for a decaying exponential input and for two time constants in the feedback loop of one and 25 times the input pulse time constant.

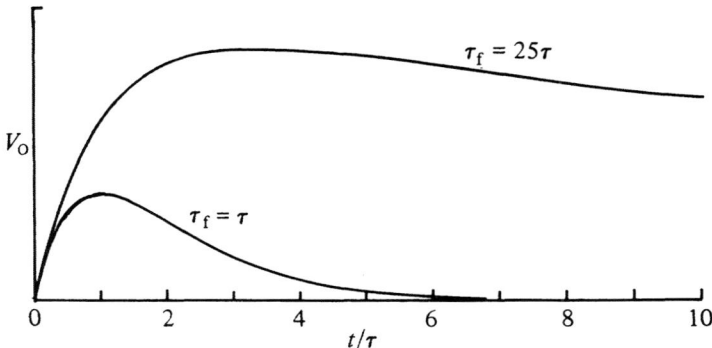

constant that is much larger than the decay constant of the input pulse. As can be seen from Fig. 7.5 the output pulse will then be an approximation to a rectangular pulse. In operation this type of pre-amplifier has an output signal that varies with time as in Fig. 7.6 with slowly decaying pulses superimposed on a steady voltage level that increases with increasing count rate. Because of the DC level this output has to be capacitively coupled to the main amplifier.

Although NIM modules of different manufacture can generally be interconnected care has to be taken to use matching pre-amplifiers and main amplifiers since the main amplifier design depends on the type of pre-amplifier used. An integrating pre-amplifier must not be used with a main amplifier designed for use with a non-shaping pre-amplifier and vice-versa or very severe pulse shape distortion will occur.

7.2.2   Main amplifiers

The purpose of a main amplifier is to amplify the small pulses from the pre-amplifier up to amplitudes of the order of volts so that the pulses can be further processed by discriminators, scalers, pulse height analysers, etc., and also to shape the pulses so that they will be compatible with the input requirements of the processing modules. Pulse shaping is achieved by integrating the detector pulse and following this with further stages of differentiation and integration. As has already been seen some pre-amplifiers carry out the first integrating stage of pulse shaping.

In order to gain an idea of how pulse shaping is achieved it is convenient to assume that the integrated input pulse has the profile of a rectangular step. Provided that the integrator time constant is

Fig. 7.6. Typical pulse sequence from an integrating pre-amplifier.

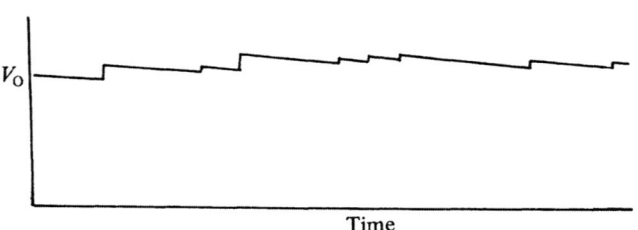

Time

much longer than the decay constant of the input pulse this is a reasonable approximation to the true behaviour. Two simple RC circuits are used for differentiation and integration and these are shown in Figs. 7.7(a) and (b) respectively. When connected together they have to be buffered by amplifiers having a very high input impedance and a very low output impedance in order to avoid loading of these circuits and consequent distortion of their functions.

Consider a step function beginning at time zero applied to the input of the differentiator. By equating voltages the following expression is obtained:

$$V_i = \frac{q}{C_1} + V_o$$

By differentiating this expression it becomes

$$\frac{dV_i}{dt} = \frac{i}{C_1} + \frac{dV_o}{dt} \tag{7.8}$$

where $i = dq/dt$. Since $i = V_o/R_1$ then equation (7.8) becomes

$$\frac{dV_o}{dt} + \frac{V_o}{R_1 C_1} = \frac{dV_i}{dt} \tag{7.9}$$

Solving equation (7.9) for a step function leads to

$$V_o = V_i \, e^{-t/R_1 C_1} \tag{7.10}$$

Now consider the integrator in Fig. 7.7(b) for which the charge on the capacitor is

$$q = C_2 V_o$$

$\therefore$

$$i = \frac{dq}{dt} = C_2 \frac{dV_o}{dt}$$

Fig. 7.7. RC circuits for (a) differentiation and (b) integration.

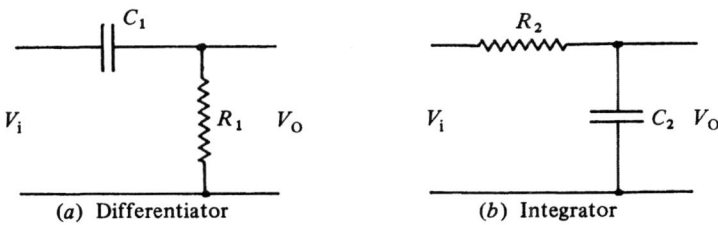

(a) Differentiator   (b) Integrator

Also $V_o + iR_2 = V_i$ and therefore

$$\frac{dV_o}{dt} + \frac{V_o}{R_2 C_2} = \frac{V_i}{R_2 C_2} \tag{7.11}$$

If now the input voltage V_i is replaced by $V_i \exp(-t/R_1 C_1)$ an equation similar in form to equation (7.5a) is obtained and this has a solution similar to equation (7.7):

$$V_o = \frac{-V_i R_1 C_1}{(R_2 C_2 - R_1 C_1)} (e^{-t/R_2 C_2} - e^{-t/R_1 C_1}) \tag{7.12}$$

If $R_1 C_1 = R_2 C_2 = RC$ then equation (7.12) has a special form:

$$V_o = V_i \frac{t}{RC} e^{-t/RC} \tag{7.12a}$$

In order to achieve the integration of the input waveform it is essential to feed the differentiator into a very high impedance and to feed the integrator from a very low impedance source. It is obvious that the circuits (7.7a and b) cannot be directly connected. An amplifier must be used to satisfy the output and input conditions as depicted in Fig. 7.8. A unity gain amplifier has been shown in order to be consistent with equation (7.12) but in general an amplifier with a higher gain would be used, thus achieving some of the necessary pulse amplification. It would also generally be an inverting amplifier.

If further stages of integration are added the pulse is lengthened but becomes more symmetrical and the long decay tail is reduced. As shown in Fig. 7.8 if the differentiation and integration time

Fig. 7.8. Step input pulse to a differentiating and integrating circuit in sequence. Note the buffering of the differentiator output by an amplifier. Output pulse is shown for equal differentiating and integrating time constants.

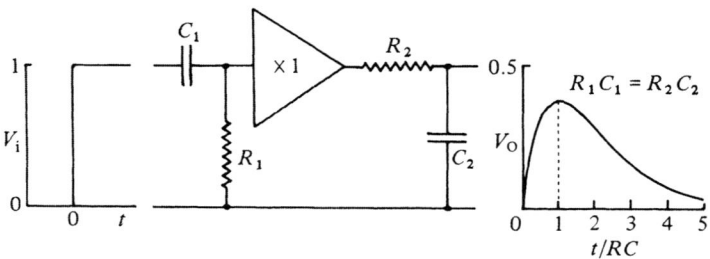

constants are equal, then for one stage of integration the peak occurs at a time $t = RC$. For a total of four successive stages of integration, all with equal time constants RC, then

$$V_o = V_i\left(\frac{t}{RC}\right)^4 e^{-t/RC} \qquad (7.13)$$

This expression peaks at $t = 4RC$ and drops to less than 1% of the peak value at a time $10RC$ after the peak. The benefits of integration are a reduction in the high frequency noise caused by the differentiator stage and the production of a pulse with a rounded top that is more suitable for analysis for peak height.

Unfortunately the input pulse from a pre-amplifier is not the ideal step function that has been assumed so far but has a slow exponential decay. (There is one special type of pre-amplifier mainly used with Si(Li) X-ray detectors that does have an infinite time constant and hence does supply a step function with a finite rise time to the main amplifier. This has an arrangement to reset the output voltage level to zero just before the output becomes saturated.) As a result of the slow decay of the step part of a real pulse the output pulse from the main amplifier tends to undershoot, as indicated in Fig. 7.9. Such undershoot could cause errors in measuring pulse height for another pulse following rapidly on the first but can be eliminated by addition of a resistor R_{pz} in parallel with the input capacitor, as is also shown in Fig. 7.9. The value of R_{pz} has to be adjusted to suit the detection system attached to the amplifier and the method of correcting pulse shape is known as *pole zero cancellation*.

Another problem that can arise with amplifiers that are AC coupled, particularly in the output stage, is the tendency for the blocking capacitor to charge up so that the heights of the pulses above the zero baseline are less than their true heights. This effect increases with increasing counting rate and is depicted in Fig. 7.10. This can be corrected by the method shown in Fig. 7.10 where S is in reality an electronic switch that is only open for a fixed period during each pulse. This correction method is known as *baseline restoration*.

An alternative and widely used method to reduce baseline shift is to differentiate the pulse twice; that is, the step input pulse is first differentiated, then integrated and then again differentiated. If equation (7.12a) is regarded as being the form of input voltage applied to equation (7.9) and all time constants are assumed to be equal then the final output voltage for an initial step input voltage of V_i is

$$V_o = V_i \frac{t}{RC} \left(1 - \frac{t}{2RC}\right) e^{-t/RC} \qquad (7.14)$$

Fig. 7.11 shows the circuit of a doubly differentiating amplifier using unity gain buffers to couple the stages. In a practical amplifier these amplifiers would probably have a gain greater than unity.

Fig. 7.9. Undershoot of unipolar pulse due to decay of input pulse and the pole-zero-cancellation circuit to eliminate this effect.

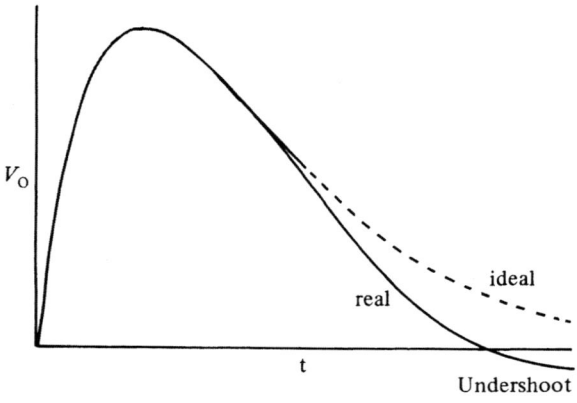

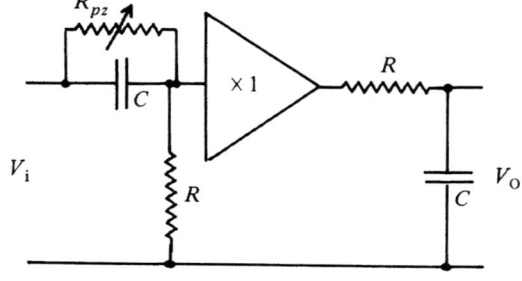

Fig. 7.10. Baseline shift at high counting rates from an AC coupled
output and a schematic baseline restoration circuit.

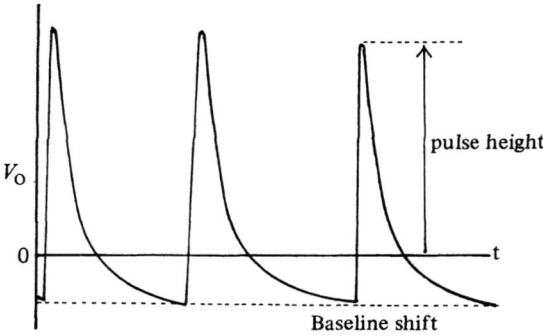

V_0

pulse height

0   t

Baseline shift

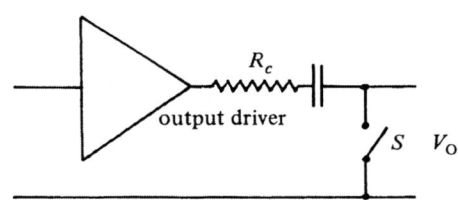

R_c

output driver

S V_0

Fig. 7.11. Double differentiation amplifier for production of a bipolar
pulse.

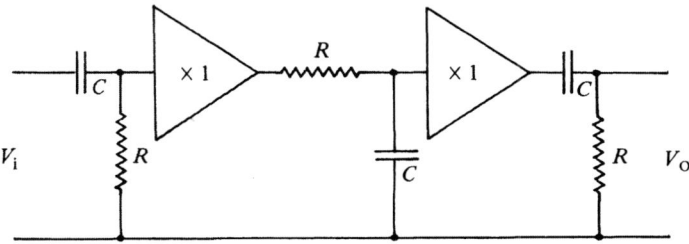

C $\times 1$ R $\times 1$ C

V_i R C R V_0

Fig. 7.12 shows the output pulse obtained with the circuit of Fig. 7.11 for a step voltage input. The undershoot introduced by the second differentiation greatly reduces baseline shift but since the areas above and below the zero volts axis are not identical (upper 46% of total, lower 54% of total) then some baseline shift can still be expected. Such pulse shaping can be useful at high counting rates but does tend to be noisier than the single stage of differentiation. Another use for such pulse shaping is timing of events since timing units are available that are triggered at the point where the pulse crosses zero.

7.3   Interconnection of units

Coaxial cable is normally used for signal transfer since it has more uniform characteristics than most twin wire cables and also has better screening against electrical interference. Many variations of the coaxial cable exist and the main characteristics of importance are the voltage breakdown and the characteristic impedance. Cable diameter is also important and must be chosen to match the connectors that have to be used at the cable ends. Characteristic impedance depends upon the series inductance and the parallel capacitance of the cable per unit length and is important for pulse and high frequency AC signals. The most common characteristic impedance of coaxial cable that is used for interconnecting units is $50\,\Omega$. Characteristic impedance can be pictured

Fig. 7.12. Bipolar output pulse from the circuit in Fig. 7.11 for a step input of unit amplitude.

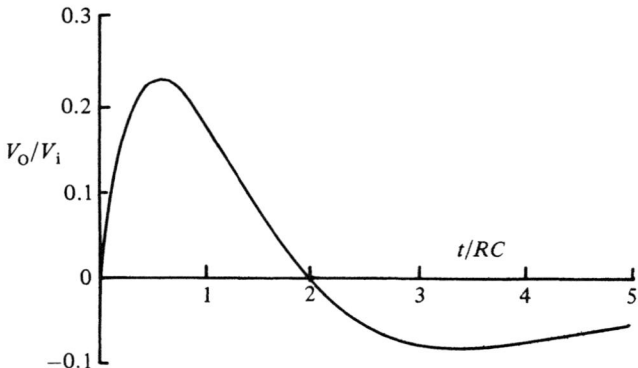

as the load that an infinite length of cable presents to an input or output. Since all cables are of finite length the cable termination is important since it may cause signals to be reflected, thus upsetting pulse shape or timing of the pulse. The velocity of signal propagation in a coaxial cable is typically about two-thirds of the velocity of light or about 2×10^8 m s^{-1} (reference should be made to manufacturers' data for exact specifications) so the signal will travel about 1 m in about 5 ns. When long cable runs are necessary then multiple reflections of pulse signals from the ends can grossly distort the information. The way of overcoming this difficulty is by correct termination of the cable ends. Consider a finite length cable of characteristic impedance 50Ω which has a 50Ω resistor connected across the inner and outer conductors at the end remote from the input. If the input signal is a step voltage of amplitude V then a current of V/50 A will flow until the leading edge of the step reaches the far end of the cable. Since the terminating resistor has the same value as the characteristic impedance then the same current will continue to flow and so there will be no reflection and the cable appears to be infinite in length, as seen from the input. Some NIM modules that require fast timing signals have input impedances of 50Ω and so the input cable is correctly terminated. Pre-amplifier outputs can often be adjusted to have an impedance equal to the characteristic impedance of the cable being used but care must be taken here since some manufacturers use a fixed output impedance of 93Ω.

It does not in general matter which end of the cable is correctly terminated and very short cables may neeed no termination, but if distorting reflections are found to be present then the user must add suitable termination. If the signal is being fed to a high impedance input $Z_i > Z_c$ then a parallel termination is used, as shown in Fig. 7.13.

Fig. 7.13. Termination of a cable connected to a unit having an input impedance greater than the characteristic impedance of the cable.

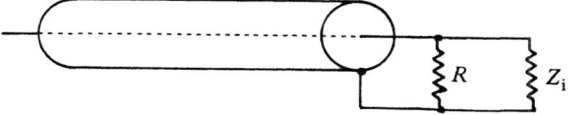

The terminating resistor has the value

$$R = \frac{Z_i - Z_c}{Z_i Z_c}$$

This type of termination is often necessary when viewing pulse shapes with the aid of a cathode ray oscilloscope that can typically have an input impedance of 1 MΩ.

If, however, the signal is being fed from a low impedance source (or to an input having an impedance less than the characteristic impedance) then a series resistor is needed, as shown in Fig. 7.14, and the value of the series resistor is given by

$$R = Z_c - Z_i$$

Note that if a cable is being fed from a low impedance source and is correctly terminated at both ends then the transmitted signal at the output end is halved in amplitude.

Care must also be taken if it is desired to split a signal so that it feeds two modules in parallel. One simple way of doing this is to use a resistive attenuator which will halve the voltage amplitude of the signal at the two outputs compared with the input. Such an attenuator is shown in Fig. 7.15 and this maintains an overall loading of Z_c at the input provided that the output cables are correctly terminated.

Connectors should be chosen to match the cable used both for size and for characteristic impedance. The BNC connector, a bayonet locking type which is very widely used, can be obtained in different versions to suit different cables, although 50Ω is the most common version. High voltage connectors are not so standard between manufacturers. Many use the SHV connector, a bayonet locking type in which the high voltage pin is well shrouded, but the MHV (or high voltage BNC) is also used. This is another bayonet

Fig. 7.14. Termination of a cable connected to a unit having an input (or output) impedance less than the characteristic impedance of the cable.

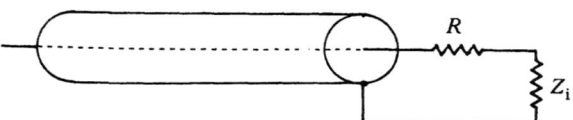

locking type that can unfortunately be made to connect with an ordinary signal BNC if excessive force is used, resulting in a poor connection and often damaged connectors. If the cable is being used solely to convey high voltage then the characteristic impedance is unimportant and any termination can be used. With gas-filled detectors and sometimes with scintillation counters the signal is taken along the high voltage cable and it may be necessary to terminate the cable to prevent signal distortion.

7.4 Pulse processing modules

7.4.1 Discriminators

One of the most useful modules used in pulse processing is the discriminator. This unit will only pass a pulse to the output provided that it is above a preset voltage level, and so it can be used to cut out small pulses that may be mixed with the inherent noise of the amplifier system. A version of the operational amplifier known as a comparator can be used for this purpose, as shown in Fig. 7.16. No feedback is used on the amplifier so that the output will saturate to a positive or zero value according to the relative signs of the inputs. (A normal operational amplifier would saturate to a positive or negative value.) It will be observed that the amplitude of the output pulse is not related to the amplitude of an accepted input pulse but is just a logic level that can be used to drive a scaler, for example. Two main logic levels are in use in nuclear pulse processors, a logic 1 TTL pulse ($+5$ V) or a negative 16 mA current signal that is mainly used for fast timing.

Fig. 7.15. Splitting of a signal to maintain the correct loading on the input cable. The output cables should be correctly terminated for proper operation.

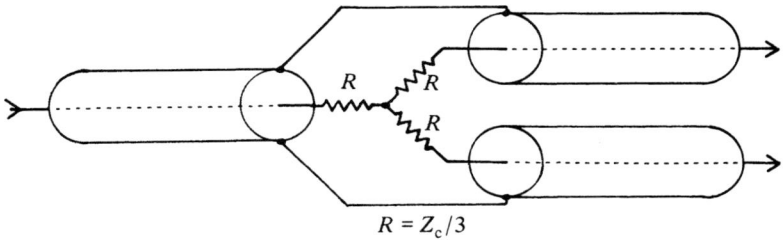

$$R = Z_c/3$$

Fig. 7.16. Use of a comparator as a discriminator.

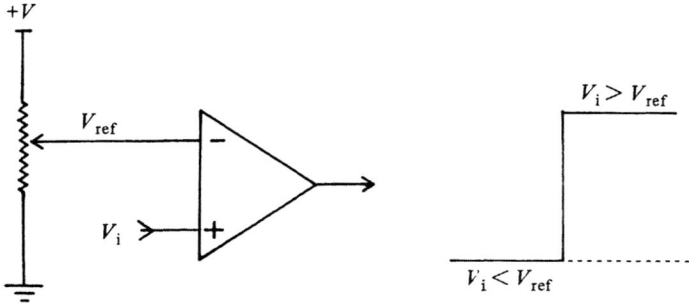

Fig. 7.17. Use of two comparators as a single channel analyser.

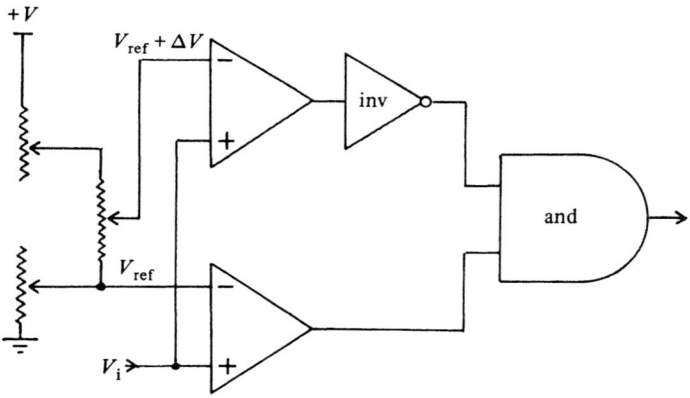

Fig. 7.18. A simple ratemeter circuit.

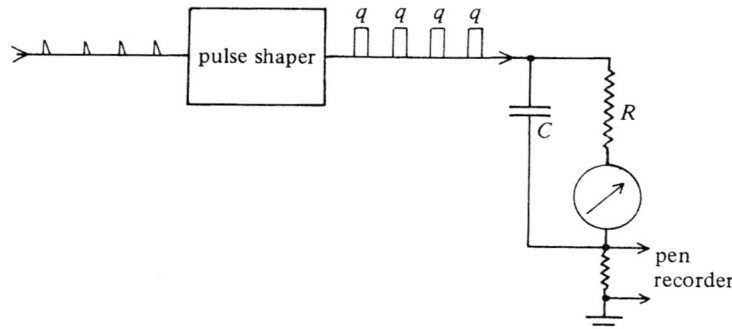

Two comparators can be used as a single channel analyser such that only pulses whose amplitudes fall between two preset limits will produce an output pulse. In Fig. 7.17 the outline circuitry of a single channel analyser is shown and it will be seen that only pulses whose amplitude falls between the two preset limits V and $V + \Delta V$ will enable the AND gate that produces the output pulse.

7.4.2 Scalers and ratemeters

Scalers are just circuits that count the number of input pulses and display the total counts recorded by some means. Most are based on 'divide by ten' integrated circuits with a direct readout of the state of the dividers by LED numerals. Since they are based on standard integrated circuits no attempt will be made to describe the detailed circuitry. Scalers generally have an input discriminator that can be adjusted to suit the incoming signal, reject noise levels, and will operate on both logic level and amplifier signals.

Ratemeters fall into two classes: digital and analogue. Digital ratemeters are rather similar to scalers in that they record counts for a preset time (for example, one second) and then will reset the scaler to zero and display the counts recorded until the next counting interval is over, repeating the sequence indefinitely. A sequence of numbers that fluctuates about the mean is therefore seen, and with some digital ratemeters it is possible to send the register contents to a printer for a permanent record. Analogue ratemeters display the counting rate by the deflection of a moving coil meter as shown in Fig. 7.18. The pulse shaper is necessary to cater for different amplitude and length input pulses that may come from a range of source impedances, and it produces a sequence of pulses of constant charge. The meter records the average current due to these standard pulses and the capacitor and resistance serve to damp out fluctuations at low counting rates. Generally a time constant of a few seconds is required to obtain a fairly steady reading at count rates of only a few per second. From this type of ratemeter it is easy to provide an analogue voltage that can be fed to a pen recorder and a similar output can be obtained from a digital ratemeter by inclusion of a digital to analogue converter.

7.4.3   *Coincidence circuits*

Sometimes more than one detector is used. For example, two or more detectors that transmit the particle, absorbing only a small fraction of its energy, may be used to show the direction of an energetic particle. Another example is where one counter surrounds another in order to indicate whether the detected event is due to a particle that has passed through both detectors, such as a cosmic ray particle, so that background counts can be much reduced. In both cases it is necessary to determine whether both detectors produce a pulse 'simultaneously' or, to be precise, within a very short time interval that is known as the *resolving time*, since electronic circuits cannot handle infinitely short pulses. The first example quoted above requires an output pulse to be generated if the two pulses are coincident within the resolving time, and is simply effected by generating short logic pulses and applying these to an AND gate. This is shown in Fig. 7.19(a) and the resolving time is nearly equal to twice the width of each input pulse, assuming that they are equal in width. It is generally necessary to shape the output pulse to a standard width since the output of the

Fig. 7.19. The use of AND gates to register (a) coincidence and (b) anti-coincidence between two pulses.

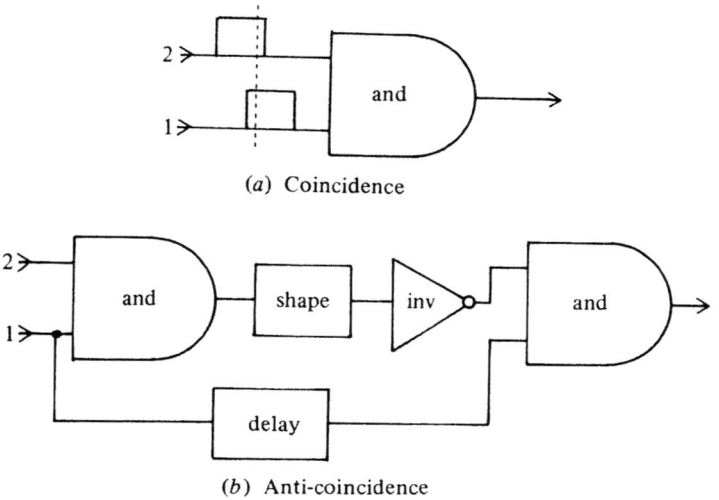

(a) Coincidence

(b) Anti-coincidence

AND gate depends on the time overlap of the two pulses. This principle can easily be extended to detect coincidence between several detectors simply by using a multi-input AND gate.

For the second case, where a guard detector is used to check for cosmic ray background pulses, no output pulse must be produced if the two detectors record simultaneously or if only the guard detector fires. This system is called an anti-coincidence circuit and the essentials of it are shown in Fig. 7.19(*b*). This uses a second AND gate whose inputs are the suitably delayed signal from the first detector and the shaped and inverted output from the AND gate that registers coincidence between detector 1 and the guard detector 2. The shaping and delay circuits are necessary to ensure correct operation of the second AND gate.

7.4.4 *Time-to-amplitude converters*

A useful way of finding the time difference between two signals is to generate a voltage pulse whose amplitude is proportional to the time difference. Time difference measurements are necessary if it is desired to find the velocity of a particle by using two transmission detectors at a known distance apart. A time-to-amplitude converter works by charging a capacitor from a constant current source (giving a linear rise in voltage with time), the charging cycle being initiated by a START pulse derived from the first detector and terminated by a STOP pulse derived from the second detector. Both these pulses generally need to be negative fast logic pulses (-16 mA) for accuracy in timing. The voltage attained by the capacitor is then presented via a buffer amplifier to the output and the capacitor is then rapidly discharged to be ready for the next pulse. Shaping is necessary for the output pulse so that it is suitable for further processing. Time intervals in the range 0–50 ns up to about 0–50 μs can be measured by such devices, the time range being decided by the size of the capacitor and the charging rate. Time within the chosen interval is then represented by a pulse with amplitude in the range 0 to $+10$ volts. This enables time differences to be recorded and displayed on a multi-channel analyser which is the subject of the next section.

7.5   Multi-channel analysers

The most useful unit for the analysis of counter data is the multi-channel analyser. A knowledge of the pulse height distribution can give information about the energies and intensities of the incident radiation. This information could be obtained for steady radiation levels (long lived sources) using a single channel analyser (see Section 7.4.1) to gradually search through the pulse height distribution. It is an extremely time consuming process, especially when high resolution is needed in the spectrum, since when examining one pulse height range all the remaining information has to be rejected.

The multi-channel analyser overcomes this difficulty by examining each pulse for height and then storing information about the number of pulses within narrow ranges of pulse height. Depending upon the analyser, a pulse range of normally 0–8 V can be split into 256, 512, 1024, 2048, and in some cases even larger, numbers of equal pulse height intervals that are known as channels. The heart of such a device is an analogue to digital converter (ADC) that produces a number proportional to the pulse height. This number is then used to select an address in a memory bank and the address contents are then incremented by one to record that pulse. Fig. 7.20 shows a schematic block outline of a multi-channel analyser

Fig. 7.20. Schematic block arrangement of a multi-channel pulse height analyser.

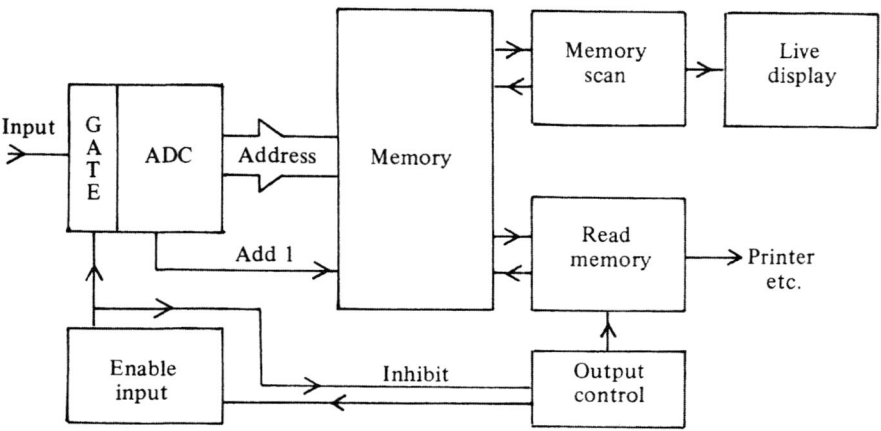

where it is assumed that amplified pulses in the range 0–8 V are available at the input. Many analysers include a built in amplifier and upper and lower level discriminators to select a desired portion of the pulse height range but these have been omitted from the diagram for clarity.

Various methods are available for analogue to digital conversion and one widely used method is the Wilkinson ADC. A capacitor is charged to the peak voltage of the incoming pulse and initially held at that level. Then it is discharged linearly from a constant current source. At the instant the discharge begins an internal gate opens to let through pulses from a high frequency oscillator, typical frequency 50 MHz, and the gate is closed at the instant when the capacitor reaches zero voltage. The pulses are counted by a binary scaler and the final number held by the scaler is used to select an address in the memory, the contents of which are then incremented by one. With fixed delays due to the various switching operations and the variable time taken to compute the address ((1/frequency) × channel number) the system has a variable dead-time per pulse, typically in the range 5–20 μs.

Another type of ADC is the successive approximation converter. Consider a poor resolution ADC for illustrative purposes that divides the 8 V pulse into eight equal steps 1 V wide and assume that the input pulse has an amplitude of 5.5 V. Initially the ADC generates the binary number 100 (decimal 4) and applies this to a digital to analogue converter (DAC) whose output (4 V) is then compared with the pulse amplitude. If the pulse is greater in amplitude than the output of the DAC then the number 100 is retained, otherwise it is set to 000. In the example chosen here it would stay at 100. The second binary digit is now set equal to 1 so that in this example the binary number becomes 110 (decimal 6) and is again applied to the DAC whose output (6 V) is again compared with the amplitude of the input pulse. In this case the pulse is less in size than the DAC output and so the second digit is reset to zero. Finally, the third binary digit is set to 1 so that the binary number becomes 101 (decimal 5) and this is again applied to the DAC whose output (5 V) is again compared with the input pulse. This time the input pulse is the greater so the binary number remains at 101 and the pulse lies within channel 5 (5–6 V). This binary number is then used to address the memory. Successive approximation ADC units have a fixed conversion time and are of course of much higher resolution than the simple 3-bit one chosen in the above example. A conversion gain of 1024 would need a 10-bit converter.

A live display of the pulse height spectrum is available via a video monitor and facilities are also included to read the spectrum (memory contents) to a printer or optionally to a graph plotter, cassette or disc recorder or to a computer for further analysis. Many models now incorporate an internal microprocessor for carrying out energy calibration, peak area and centroid determination, background subtraction and peak identification.

Further reading

Martin Hartley Jones, *A Practical Introduction to Electronic Circuits*, CUP, 1977.

D. A. Fraser, *The Physics of Semiconductor Devices*, 3rd edn, OUP, 1983.

8

....

Radiation doses and their measurement

8.1 Introduction

The common units used to express the energies of ionising radiations (keV, MeV) tend to disguise what a small amount of energy they represent in mechanical or heat units: $1 \text{ keV} = 1.6 \times 10^{-16} \text{ J}$ and $1 \text{ MeV} = 1.6 \times 10^{-13} \text{ J}$. The energy deposited by ionising radiation that is needed to damage a living cell need only be of the order of 10^{-16} J–10^{-17} J if it is received at the time of cell division. Radiation doses that can have serious effects on living matter are generally so low in energy terms that they cannot be detected by the physical senses. Detectors that respond to ionisation are therefore necessary to detect and measure ionising radiations. Before describing some of the common detectors used for dose and dose rate measurement it will be necessary to discuss the commonly used units of dose and dose rate and to relate these to fluxes of common particles and of photons.

8.2 Units of dose

The original dose unit, the *Roentgen*, is strictly not a unit of dose but of exposure and is defined as that quantity of radiation that produces 100 esu (3.33×10^{-6} C) of charge of either sign in 1 cm^3 of dry air at 760 mm pressure and at 15 °C. It is more relevant to express dose in terms of energy absorbed since this will be more directly related to tissue damage. An exposure of one Roentgen deposits 93 erg per gramme of air ($9.3 \times 10^{-3} \text{ J kg}^{-1}$ of air) and the unit of absorbed dose known as the *rad* (*roentgen absorbed dose*) was defined to be very close to this value, 100 erg absorbed per gramme of absorber (10^{-2} J kg^{-1}). In an attempt to rationalise the units the rad has been replaced by the *gray* (Gy) where one gray is one joule absorbed per kilogramme and is equal to 100 rad.

X- and gamma-radiation, which produce fast electrons, together with beta-particles produce the lowest ionisation density, leaving of the order of 10 ionisations in a typical size of cell. Heavy charged particles such as alpha-particles or protons (resulting from fast neutrons colliding with hydrogen nuclei) have a much higher ionisation density along their path than electrons and can leave the order of 100 ionisations in a typical size of cell. As a result the cell is more likely to suffer damage. The relative damaging power of different ionising radiations is described by a *quality factor* (QF) that is, by definition, 1 for X- and gamma-rays and has a value of about 7 for protons and about 20 for alpha-particles. A unit of biological dose is therefore necessary that is a measure of the actual damage received and is the absorbed energy dose times the quality factor:

Biological dose in *rem*   = (dose in rad) × QF
Biological dose in *sievert* (Sv) = (dose in Gy) × QF

This difference in damaging power for the same absorbed energy dose causes difficulties with ionisation detectors in mixed radiation fields since ionisation is a measure of the energy absorbed.

8.3   Permissible dose limits

Living cells are most sensitive to radiation damage when they are undergoing cell division or mitosis, for it is at this time that the genetic material in the cells (the chromosomes) is being replicated and consequently is most prone to damage since it has to acquire new 'building block' molecules from the contents of the cell. Mitosis rates vary enormously over different parts of the body but for whole-body exposure the critical organs are the bone marrow and the intestines. Large radiation doses received over a short period (acute doses) can cause reduction in blood cell production or have gastro-intestinal effects. A fall in the leucocyte (white blood cell) population starts to become noticeable at an acute dose of 0.25 Sv and gastro-intestinal effects at an acute dose of 1 Sv. For an acute dose of 4–5 Sv the survival rate of the exposed group would only be 50%, the mean survival time being 30 days. This is the median lethal dose and death results from secondary infections. For an acute dose in the range 8–10 Sv death is still due to secondary infections but the survival rate drops to zero without specialist medical attention. It is worth noting that for a median lethal dose of X- or gamma-radiation (QF = 1) the amount of energy absorbed in the body is only sufficient to raise the temperature of the body by 0.001 °C.

Under normal working conditions classified radiation workers (those having regular medical checks and continuous monitoring of received dose) are only permitted an *annual dose* of 50 mSv (5 rem) with a normal restriction of not more than 30 mSv (3 rem) in any 13 week period. This is based not on the acute dose effects already mentioned but is designed to minimise the risk of developing cancers such as leukaemia in later years or of acquiring genetic damage that could be passed to future generations. The annual dose can be used to define a maximum continuous working level of dose rate. If a working year of 2000 hours is assumed then the maximum average dose rate during working hours will be 25 μSv h^{-1} (2.5 m rem h^{-1}). This is then a guideline for the maximum dose rate within a working environment, assuming that the source of radiation is always present. Higher dose rates can be tolerated for shorter times provided that the restrictions on quarterly and annual dose are not exceeded. For this reason it is misleading and unnecessarily worrying to refer to a dose rate of 25 μSv h^{-1} as the maximum permissible level. For whole-body dose only the penetrating radiations, higher energy X-rays, gamma-rays and neutrons are of importance. Alpha-particles are completely stopped by a single sheet of ordinary aluminium cooking foil and about 2 mm of aluminium or 5 mm of perspex is generally adequate to absorb all beta radiation. It is therefore easy to provide shielding against these less penetrating radiations. Nevertheless, instruments to measure the presence of such radiations are still necessary since unshielded beta-particle sources can cause damage to exposed parts of the body, particularly the eyes. In addition, contamination of working surfaces in radiochemical laboratories is more easily found by detecting the alpha- or beta-radiation instead of the associated gamma-radiation. Decontamination of such surfaces is necessary to avoid accidental intake of radioactive materials into the body which can be far more hazardous than exposure to external radiations.

Returning to external sources of radiation, Fig. 8.1 shows the gamma-ray flux as a function of gamma-ray energy that gives a dose rate of 25 μSv h^{-1}, and Fig. 8.4 shows similar information for neutrons. Gamma-radiation fluxes are such that for a constant dose rate the product of flux times energy is approximately constant over a wide energy range and this leads to a useful simple guide for estimating gamma-ray dose rates at a known distance from an

Fig. 8.1. Gamma-ray flux for a dose rate of 25 μSv h^{-1} (2.5 mrem h^{-1}) as a function of gamma-ray energy. (Based on gamma free in air tissue Kerma data in ORNL/TM-4840 (1977).)

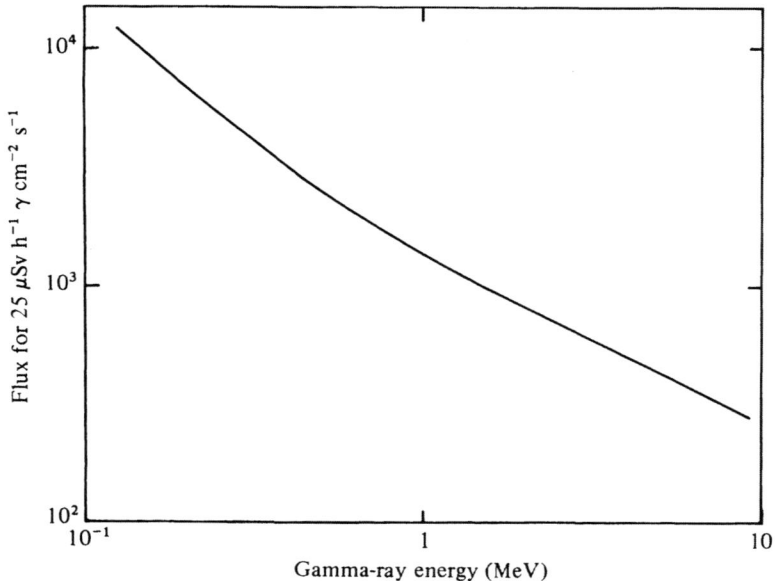

unshielded source that can effectively be regarded as a point source. At a distance of r metres from an unshielded source of strength C millicuries which emits a *total* of E MeV of gamma-rays per disintegration:

$$\text{dose rate} = 0.5\ CE/r^2 \text{ mrem h}^{-1} \qquad (8.1)$$

Note that a source strength of 1 mCi is defined as a disintegration rate of 3.7×10^7 s^{-1} or 3.7×10^7 Bq where 1 Bq (*becquerel*) is a disintegration rate of one per second and is now the preferred unit of source strength. An equivalent to equation (8.1) with the source strength expressed as B becquerel is

$$\text{Dose rate} = 1.35 \times 10^{-7}\ BE/r^2 \ \mu\text{Sv h}^{-1} \qquad (8.1a)$$

For a source emitting gamma-rays of 1 MeV reference to Fig. 8.1 shows that the gamma-ray flux giving a body tissue dose of 25 μSv h^{-1}is about 1500 cm^{-2}s^{-1}. Electronic survey instruments should be capable of measuring dose rates down to about 1 μSv h^{-1} which is

Fig. 8.2. Geiger–Mueller survey instrument. (Reproduced by permission of Mini Instruments Ltd, Burnham on Crouch.)

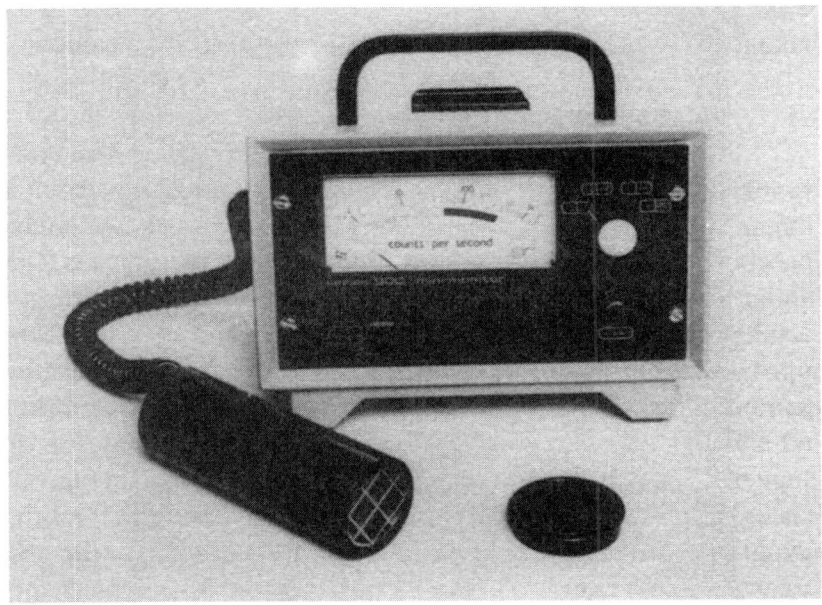

about twice the typical average value for natural background, a figure which varies with geographical and geological location and height. They should also be capable of recording dose rates in excess of 1 mSv h^{-1} to give warning of dangerous conditions.

8.4   Geiger–Mueller counter based survey instruments

Since the pulses from a Geiger–Mueller counter do not require amplification a relatively simple, light and portable battery operated survey meter can be made using a ratemeter analogue display. Fig. 8.2 shows such an instrument and, depending upon the particular Geiger–Mueller tube that is fitted, it will typically show a counting rate of about 100 s^{-1} in a gamma-ray field that gives a dose rate of 25 μSv h^{-1}. In addition, if the Geiger–Mueller counter has a thin mica end window then the system will respond to less penetrating radiations with a detection efficiency of about 50% for alpha- and beta-sources placed close to the end window, so the instrument is then suitable for checking contamination levels. It is less efficient at detecting low energy X-rays: only about 5% of 6 keV (Fe K$_{\alpha}$) X-rays incident on the end window in a collimated beam are recorded. It is, nevertheless, adequate for checking for leakage of radiation from X-ray sets and crystallographic cameras.

8.5   Ion chamber survey meters

As was seen in Section 8.2 for the definition of dose units, the ionisation produced in air, and hence the ionisation current in a mean current ionisation chamber, is closely related to the energy absorbed dose in air. The mean atomic weight of soft tissue is fairly similar to that of air so the total ionisation produced in air can also be taken as representative of the tissue absorbed dose to a close approximation. Portable instruments having an ionisation chamber filled with dry air with a sensitive volume of about 0.5 litre and a window area of about 100 cm^2 can record dose rates in the range 0.1–300 μSv h^{-1} and such an instrument is shown in Fig. 8.3. It is usual to cover the thin aluminium window (8 mg cm^{-2}) with a bakelite cover 2 mm thick for more accurate recording of the dose rate from higher energy gamma-radiation (in the MeV range) since the build up of secondary scattered gamma-rays is not as effective in the air-filled ionisation chamber as it is deep into soft tissue. A gamma-ray energy range of about 10 keV–2 MeV can be covered to an accuracy of better than ±20% in measured dose rate. It is an indication of the progress in electronics that such a small portable mean current ionisation chamber can be made to record reliably. A further feature incorporated is the ability to integrate the signal to obtain total dose. Generally, similar instruments having a small

Fig. 8.3. Ion chamber survey meter type PDM 1 for dose rate and dose measurement of β and γ. (Reproduced by permission of Nuclear Enterprises Ltd, Edinburgh.)

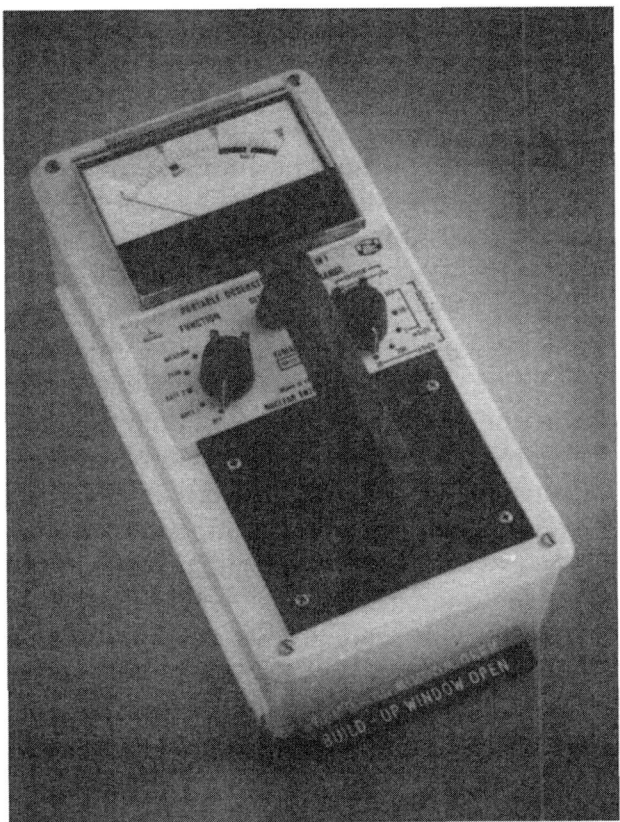

ionisation chamber on the end of a long cable are available for surveying, in detail, more intense radiation fields for medical or biological purposes.

8.6   Contamination meters

The thin end window Geiger–Mueller counter can be used as a contamination detector, as was noted in Section 8.4, but with an end window area of only about 5 cm^2 it can be time consuming, making surveys of large areas, and it is also easy not to cover the surface thoroughly. Larger area probes are specifically manufactured for the purpose of contamination measuring and one widely used form is a type of scintillation counter. For beta-particle contamination measurements a sheet of plastic scintillator of area 50 cm^2 and thickness 3 mm is suitable. This has to be covered with a sheet of thin aluminium, typical thickness 5 mg cm^{-2} to exclude light and any alpha-particles. Any detector for alpha-particles must be thick enough to stop the alpha-particles yet thin enough to have a very low response to other radiations. A thickness of about 5 mg cm^{-2} is suitable and this could be a thin layer of plastic scintillator or of silver activated zinc sulphide deposited on a sheet of clear plastic. A very thin light-tight window is needed to cover the scintillator to exclude light and yet allow the alpha-particles to enter, and this is usually made of two layers of very thin Melinex film, each layer being vacuum coated with aluminium.

A combination probe that is sensitive to both alpha- and beta-particles is made by coating the thicker plastic scintillator used for beta-particle detection with a thin ZnS(Ag) layer on the entrance window side and using the same type of thin window that is used in the alpha-particles probe. A large light pulse is produced by alpha-particles in comparison with beta-particles since the ZnS(Ag) has a high light output compared with NE102A (see Table 5.1), and the alpha-particles also have higher energies than the beta-particles so it is easy to distinguish between the two types of radiation by means of the pulse heights. The associated electronics, therefore, has to include a single channel analyser set to record the beta-particle pulse height range and a discriminator with a threshold set above the maximum pulse height to be found for beta-particles to record the alpha-particle pulses.

Maximum permissible contamination levels for alpha-emitters are 4 Bq cm^{-2} for plant, apparatus and equipment but only 0.4 Bq cm^{-2} for skin and personal clothing. For non-alpha-emitting nuclides the corresponding figures are 40 Bq cm^{-2} and 4 Bq cm^{-2}. It has to be remembered that a large surface area detector placed very close to a contaminated surface will detect less than half the emitted radiation since the radiation will be emitted isotropically, so to obtain a realistic estimate of the true disintegration rate of the contamination the measured counting rate must be multiplied by a factor of about three even if the detector is 100% efficient for radiation that strikes it. This multiplication factor should if possible be checked with a calibrated source.

8.7   Neutron survey meters

If the neutron field consists solely of thermalised neutrons (most probable energy kT or $\frac{1}{40}$ eV at room temperature) then a good measure of the thermal neutron flux can be obtained by using a BF$_3$ proportional counter. Such a detector having an active volume 20 mm in diameter and 120 mm long and filled with ^{10}BF$_3$ to a pressure of half an atmosphere has a counting rate of three counts per second in a thermal neutron flux of 1 n cm^{-2} s^{-1} and will also have a very low response to gamma-radiation provided that the small pulses due to gamma-radiation are cut out with a discriminator. This detector will not respond significantly to fast neutrons either because of the low cross-section for the reaction at higher energies and the low energy of recoil particles resulting from elastic scattering which also has a low cross-section compared with that of ^{10}B at thermal energies.

In general, whether the neutrons arise from accelerators, reactors or radioactive sources there will be a very wide range of neutron energies present, ranging from thermal to several MeV and any detector must be capable of giving an indication of the biological dose (in sievert or rem) over the whole energy range. As shown in Fig. 8.4 the flux for a constant dose rate varies widely over the possible range of neutron energies. The usual solution to this problem is to slow the fast neutrons down in a hydrogenous moderator and count the thermalised neutrons with, for example, a BF_3 proportional counter. Since for the same dose rate thermal neutron fluxes are about 30 times those in the MeV range it is also necessary to attenuate the thermal neutron component of the incident flux before it reaches the detector. One design of wide energy range neutron dose rate meter is shown in Fig. 8.5. The main feature is a polythene cylinder 216 mm diameter and 250 mm long which has a BF_3 detector at the centre. At roughly half-radius a perforated cadmium shield is included to reduce the intensity of incident thermal neutrons. Fast neutrons will mainly be slowed down inside the cadmium shield and so the attenuation effect on slowed down fast neutrons will be small. This instrument is capable of indicating the biological dose over the range thermal to 10 MeV within ±20%, but for higher energy neutrons the response falls off markedly since the slowing down distance for neutrons increases rapidly with energy. A larger moderating block would improve the

Fig. 8.4. Neutron flux for a dose rate of 25 μSv h^{-1} (2.5 mrem h^{-1}) as a function of neutron energy. (Based on Snyder-Auxier neutron tissue dose data in ORNL/TM-4840 (1977).)

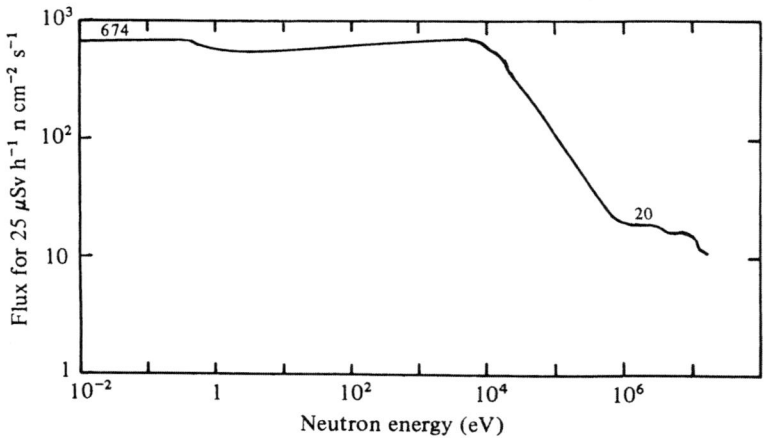

fast neutron energy range but the weight of the instrument would then become excessive. An analogue and digital indication of dose rate is provided in this instrument and dose rates up to $100 \, \text{mSv} \, \text{h}^{-1}$ ($10 \, \text{rem} \, \text{h}^{-1}$) can be measured.

8.8 Pocket ionisation chamber

So far all the instruments described have been electronically operated but a useful range of instruments exists that are very compact, can be worn on the person and do not require electrical supplies for their operation, although with two of the instruments to be described there has to be extra equipment to read the recorded doses. The first is the pocket ionisation chamber, which only requires a simple charging unit to prepare the instrument for use, readout being purely visual. The pocket ionisation chamber is based on the gold leaf electroscope that was widely used as a detector in the early days of research into radioactive materials.

Fig. 8.5. Neutron monitor type NM 2 (range 10 MeV to thermal). (Reproduced by permission of Nuclear Enterprises Ltd, Edinburgh.)

Fig. 8.6 shows the essential features of the device. Unlike the gold leaf electroscope, in which the electrostatic repulsion of the leaf from its support is counteracted by the gravitational effect on the leaf, the deflecting element has a restoring force due to its own rigidity and comprises a loop of fine quartz fibre that has been coated with a thin layer of gold by vacuum deposition in order to make it an electrical conductor. A built-in fixed focus microscope produces an image of the end of the loop of quartz fibre on a scale (generally calibrated in roentgen) and an external charging unit to apply a voltage that can be varied, and of typical value 200 V, is all the additional equipment that is needed. The fully charged fibre is arranged to coincide with zero dose on the scale and the drop in voltage on the fibre is proportional to the integrated dose received. As the voltage falls the image of the end of the fibre moves across the scale, thus showing the total dose received. Since the readout is simply visual it is possible to keep a frequent check on the received dose, so it is useful for workers who are in high radiation fields for short periods. Such an instrument will only respond significantly to

Fig. 8.6. Pocket ionisation chamber: 1. ionisation chamber section; 2. conducting plastic wall (mean atomic weight similar to air); 3. gold coated quartz fibre and support; 4. insulator, transparent polystyrene; 5. metal bellows; 6. charging pin; 7. glass insulator; 8. objective lens; 9. graticule, calibrated in dose units; 10. eyepiece lenses; 11. metal case (aluminium or brass).

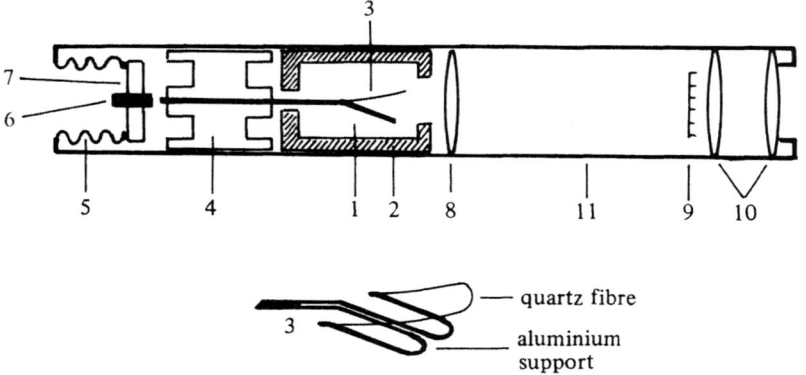

gamma-radiation. Attempts have been made to measure fast neutron dose by incorporating a hydrogenous liner in the ionisation chamber in place of the air equivalent liner that is normally used, but the device records total ionisation due to both fast neutrons (via the knock-on protons from the liner) and gamma-rays, and since these radiations have different quality factors it does not directly indicate the biological dose. Biological dose, therefore, has to be found by subtracting the gamma-ray dose as measured by a second instrument and then multiplying the remaining dose by the quality factor for neutrons. Ways have been suggested for producing a direct reading device by incorporating two ionisation chambers within one case, one gamma-ray sensitive only and the other gamma-ray and fast neutron sensitive. By coupling these two chambers with a suitably sized capacitor it is possible to make the instrument display only neutron dose.

The basic gamma-ray instrument is available in a wide range of sensitivities with dose ranges 0–200 mr (0 to 2 mGy) up to a maximum reading of 500 r (5 Gy). The higher range (less sensitive) instruments are produced by the addition of a very low leakage capacitor in parallel with the ionisation chamber in order to increase the stored charge.

The remaining two devices to be described give no direct indication to the user of received dose and require additional processing equipment in order to determine the recorded dose. They are the film badge dosimeter and the thermoluminescent dosimeter.

8.9   Film badge dosimeter

The film badge dosimeter is currently the most widely used device for personal dosimetry and responds to electromagnetic radiations from low energy X-rays to high energy gamma-rays as well as beta-radiation. It is based on the production of a latent image in a photographic film by ionising radiations, and originally the film was the familiar dental X-ray film. Nowadays it is a specially manufactured film of similar size which is coated on both sides with emulsions of different sensitivities. In order to distinguish between different types and energies of radiations it must be used in a special holder whose features are shown in Fig. 8.7. The incident radiation is filtered by various thicknesses of plastic and by metal filters and the various regions numbered in Fig. 8.7 have the following purposes:

(1) The central window region responds to all radiation that can penetrate the light tight wrapping and the film identity number also shows in the window. This identity number is

Fig. 8.7. Film badge holder: 1. open window (lets through X, β, γ); 2. plastic 50 mgm cm^{-2} (lets through X, γ and partially stops low energy β); 3. plastic 300 mgm cm^{-2} (lets through X, γ); 4. duralumin 1 mm thick (lets through γ, higher energy X-rays); 5. tin 0.7 mm + lead 0.3 mm (lets through γ-rays); 6. cadmium 0.7 mm + lead 0.3 mm (lets through γ-rays and responds to slow neutrons by production of capture γ-rays from cadmium).

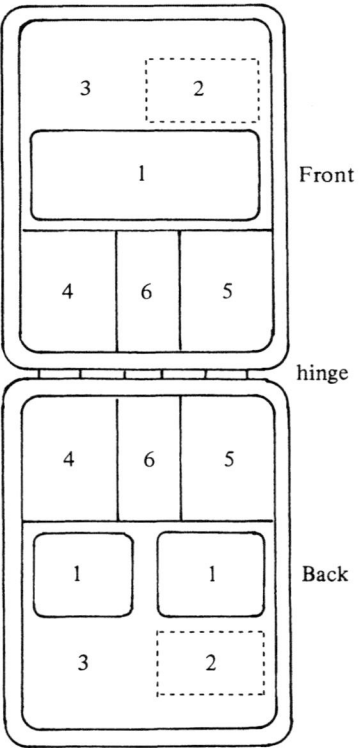

also embossed on the film by pressure and so produces a developable image for identification after processing.

(2) The thin plastic area (50 mg cm^{-2}) lets through X- and gamma-radiation and partially stops low energy beta-particles.

(3) The thick plastic area (300 mg cm^{-2}) only passes X- and gamma-radiation.

(4) The 1 mm duralumin filter lets through gamma-radiation and only higher energy X-radiation.

(5) This is a sandwich filter composed of 0.7 mm of tin and 0.3 mm of lead and only passes gamma-radiation (that is strictly radiation, X- or gamma-, with energy in excess of about 100 keV).

(6) This is also a sandwich filter composed of 0.7 mm of cadmium and 0.3 mm of lead. Since cadmium ($Z = 48$) and tin ($Z = 50$) are close together in the periodic table the absorption characteristics of this filter will be close to those of the previous filter and it will also only pass gamma-radiation. The presence of cadmium renders this section responsive to slow neutrons by the reaction

$$^{113}\text{Cd} + \text{n} \rightarrow {}^{114}\text{Cd} + \gamma$$

which has a high cross-section (2500 b) for neutrons of energy less than 0.4 eV. The emitted gamma-radiation following slow neutron capture will interact with the film beneath this filter and so will indicate the presence and dose of slow neutrons.

Fig. 8.8 shows schematically the response of a film to different types of radiation. In general the user will have been exposed to a variety of radiation types and energies and so the different regions of the film will show different optical densities after processing. Determination of total dose is carried out by successive subtraction from the different areas, starting with regions (5) and (6) and working backwards to the open window region. This process is generally automated and is linked to a computer for dose determination and production of records.

The useful dose range of the film badge dosimeter is decided by the dose/optical density characteristics of the emulsion. If light of intensity I_0 is incident on the processed film and emerges with a lower intensity I then I/I_0 is the fraction transmitted and is termed the opacity. Optical density is defined as $\log_{10}(I/I_0)$ and has a minimum value of zero for full transmission and typically a maximum density of 3 ($\frac{1}{1000}$ of the incident light is transmitted by a completely 'black' film). The dose/optical density relationship has the form shown in Fig. 8.9. An upper dose limit of 50–100 mSv is obtained and higher doses cannot be determined since they will not blacken the film any further. To overcome this restriction the film has a second emulsion on the reverse side which remains unaffected over most of the range of the sensitive emulsion. It therefore takes

Fig. 8.8. Schematic response of film badge exposed in the holder depicted in Fig. 8.4 to different radiation: 1. β ^{204}Tl E_{max} 0.78 MeV; 2. β ^{90}Y E_{max} 2.25 MeV; 3. X 20 keV; 4. X 150 keV; 5. γ ^{60}Co 1.17, 1.33 MeV; 6. thermal neutrons.

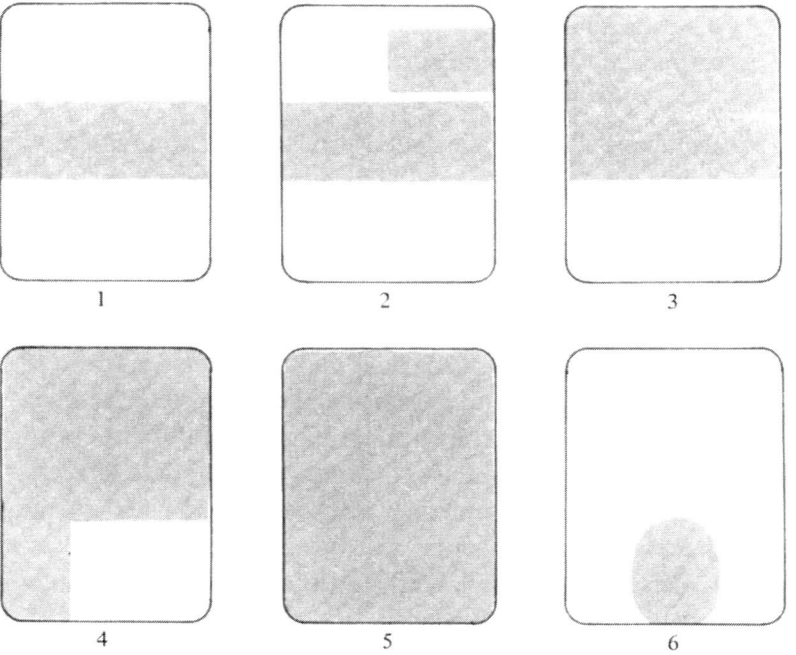

over from the blackened emulsion at higher doses and can be examined after removal of the completely blackened emulsion.

Since there are inevitably batch differences in film manufacture and processing solutions will vary according to the degree of use and their temperatures, it is normal to retain some films from a batch, expose them to standard doses and process them with the rest of the batch. This then calibrates the dose/optical density characteristic for each batch.

Films are normally worn for four weeks and a further two to three weeks is required for collection, processing and analysis of films, so if there is a risk of high radiation doses a second instrument such as a pocket ionisation chamber should also be used and surveys should be made with portable electronic instruments to give immediate warning of hazardous conditions.

Workers who are exposed to fast neutrons need to use a second type of film badge in addition to the common type. The holder contains lead on the side worn away from the body to minimise fogging by X-rays and has a tissue equivalent plastic on the side worn towards the body (this is a plastic with a similar mean atomic

Fig. 8.9. Optical density of developed film as a function of dose.

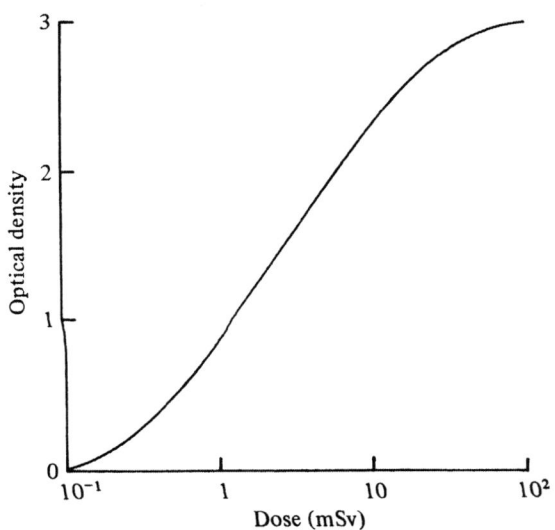

weight to soft tissue). A different type of film is used which has a thick, high density very fine grained emulsion of the type used in cosmic ray research and this will record recoil proton tracks from the plastic backing of the film. Two films are enclosed within one light-tight envelope. One has a thin plastic backing and the other a thicker plastic backing and thicker emulsion. In both cases the emulsion side is worn towards the body. This combination enables neutron energies to be obtained from the lengths of the recoil proton tracks and the intensity from the number of tracks. Analysis is, however, tedious and expensive since each processed film has to be examined under a microscope by an operator trained to interpret the proton tracks.

8.10   Thermoluminescent dosimeter

The thermoluminscent dosimeter is used mainly for the measurement of gamma-radiation dose and comprises a pellet of, for example, CaF_2 plus a few percent of manganese ($CaF_2 \cdot Mn$) or $CaSO_4$ or LiF. When these materials are irradiated electrons and holes are produced and instead of recombining the electrons are trapped in so-called trapping centres present in the forbidden gap. The purpose of the manganese in $CaF_2 \cdot Mn$ is to produce suitable traps. All thermoluminescent dosimeter materials are chosen for a high density of trapping centres and in addition they must release visible light on heating. Heating releases the energy stored in the trapping centres (the electrons are released to recombine with the holes in the valence band). A high proportion of the stored energy must be released as light if the material is to be useful for dosimetry. The total light released on heating is proportional to the absorbed dose over a very wide dose range. A reader is required that takes the detector through a carefully controlled and reproducible heating cycle up to a maximum temperature of about 300 °C, the exact cycle and temperature reached being dependent on the detecting material used. A photomultiplier tube detects the visible light released and its output is fed to an integrator for total dose determination. Unlike the film badge dosimeter the thermoluminescent dosimeter is reusable after reading since the trapped charge carriers have been returned to their lowest energy states and the detector has been 'wiped clean'.

The light output is proportional to total dose over a wide dose range. For example, LiF which, being a low atomic weight material has absorption characteristics generally similar to soft tissue, has a minimum sensitivity of about 100 μSv (10 mrem) and is linear up to a dose of about 10 Sv (1000 rem). Because of the presence of the isotope ^{6}Li this substance will also be sensitive to slow neutrons through the reaction

$$^6\text{Li} + \text{n} \rightarrow \alpha + {}^3\text{He}$$

but the slow neutron sensitivity can be eliminated by using the separated isotope ^{7}Li in the form ^{7}LiF, a solution which is expensive due to the cost of isotopic separation.

Further reading

J. J. Fitzgerald, G. L. Brownell & F. J. Mahoney, *Mathematical Theory of Radiation Dosimetry*, Gordon & Breach, 1967.

A. Martin & S. A. Harbison, *An Introduction to Radiation Protection*, Chapman & Hall, 1972.

Index

... .